Zhenhao Zhang

Nanoscale investigation of potential distribution in operating Cu(In,Ga)Se$_2$ thin-film solar cells

Nanoscale investigation of potential distribution in operating Cu(In,Ga)Se$_2$ thin-film solar cells

by
Zhenhao Zhang

KIT Scientific Publishing

Dissertation, Karlsruher Institut für Technologie (KIT)
Fakultät für Elektrotechnik und Informationstechnik, 2012

Impressum

Karlsruher Institut für Technologie (KIT)
KIT Scientific Publishing
Straße am Forum 2
D-76131 Karlsruhe
www.ksp.kit.edu

KIT – Universität des Landes Baden-Württemberg und
nationales Forschungszentrum in der Helmholtz-Gemeinschaft

KIT Scientific Publishing 2013
Print on Demand

ISBN 978-3-86644-978-7

Nanoscale investigation of potential distribution in operating Cu(In,Ga)Se$_2$ thin-film solar cells

zur Erlangung des akademischen Grades eines

DOKTOR-INGENIEURS

von der Falkutät für
Elektrotechnik und Informationstechnik
des Karlsruher Instituts für Technologie (KIT)
genehmigte

Dissertation

von

Dipl.-Ing. Zhenhao Zhang
geb. in Shandong/China

Tag der mündlichen Prüfung: 14.12.2012
Hauptreferent: Prof. Dr.-Ing. Michael Powalla
Korreferent: Prof. Dr. rer. nat. Clemens Heske

Publications

Journal publications

- **Z-H. Zhang**, M. Hetterich, U. Lemmer, M. Powalla and H. Hölscher, "Cross sections of operating $Cu(In,Ga)Se_2$ thin-film solar cells under defined white light illumination analyzed by Kelvin probe force microscopy", Appl. Phys. Lett., **102**, 203903 (2013).

- **Z-H. Zhang**, W. Witte, O. Kiowski, U. Lemmer, M. Powalla and H. Hölscher, "Influence of the Ga content on the optical and electrical properties of $CuIn_{1-x}Ga_xSe_2$ thin-film solar cells", IEEE J. Photovoltaics, DOI: 10.1109/JPHOTOV.2012.2228297.

- **Z-H. Zhang**, X-C. Tang, O. Kiowski, M. Hetterich, U. Lemmer, M. Powalla and H. Hölscher, "Reevaluation of the beneficial effect of $Cu(In,Ga)Se_2$ grain boundaries using Kelvin probe force microscopy", Appl. Phys. Lett., **100**, 203903 (2012).

- **Z-H. Zhang**, X-C. Tang, U. Lemmer, W. Witte, O. Kiowski, M. Powalla and H. Hölscher, "Analysis of untreated cross sections of $Cu(In,Ga)Se_2$ thin-film solar cells with varying Ga content using Kelvin probe force microscopy", Appl. Phys. Lett., **99**, 042111 (2011).

- F. Nickel, M. Reinhard, **Z-H. Zhang**, A. Pütz, S. W. Kettlitz and U. Lemmer, "Solution processed sodium chloride interlayers for efficient electron extraction from polymer solar cells", Appl. Phys. Lett., **101**, 053309 (2012).

- M. Reinhard, J. Hanisch, **Z-H. Zhang**, E. Ahlswede, A. Colsmann and U. Lemmer, "Inverted organic solar cells comprising a solution-processed cesium fluoride interlayer", Appl. Phys. Lett., **98**, 053303 (2011).

- A. Slobodskyy, D. Wang, C. Kübel, **Z-H. Zhang**, E. Müller, D. Gerthsen, M. Powalla and U. Lemmer, "Optical and charge transport properties of a $CuIn_{1-x}Ga_xSe_2$ solar cell cross section", Phys. Rev. B, **82**, 201105 (2010).

Conference oral presentations

- **Z-H. Zhang**, W. Witte, O. Kiowski, U. Lemmer, H. Hölscher and M. Powalla, "Influence of the Ga content on the optical and electrical properties of $CuIn_{1-x}Ga_xSe_2$ thin-film solar cells", IEEE PVSC, Austin (June 2012).

- **Z-H. Zhang**, V. Haug, I. Klugius, M. Reinhard, A. Quintilla, T. Magorian Friedlmeier, E. Ahlswede, A. Colsmann, M. Powalla and U. Lemmer, "Impact of the reduction of the selenization duration in a sequential process on the properties of $Cu(In,Ga)Se_2$ thin-film solar cells", EMRS Bilateral Energy Conference, Nice (May 2011).

Conference posters

- **Z-H. Zhang**, X-C. Tang, U. Lemmer, O. Kiowski, M. Powalla and H. Hölscher, "Comparison of the potential distributions at grain boundaries on the surface and in the depth of the absorber in $Cu(In,Ga)Se_2$ thin-film solar cells", IEEE PVSC, Seattle (June 2011).

- **Z-H. Zhang**, M. P. Heinrich, A. Slobodskyy, A. Pütz, M. Reinhard, A. Colsmann, U. Lemmer and M. Powalla, "Investigation of defect distributions at the BPhen:Li/CIGS-interface in hybrid solar cells", EMRS Spring Meeting, Strasbourg (June 2010).

Contents

1 Introduction

Today, energy plays a central role in international relations and in the daily life of civilizations. In the last two hundred years, the world consumption of primary energy has explosively increased and reached 12,274.6 Mtoe (Million tonnes oil equivalent) in 2011 [1]. Among different kinds of energy resources, fossil fuels dominate with a market share of 87% [1]. However, fossil fuels are limited. Their spiking prices in 2008 and the European gas crises in 2007 and early 2009 showed the necessity to diversify energy supplies and highlighted the importance of local renewable energy resources such as photovoltaics. Indeed, years before the most recent energy crisis photovoltaics already became one of the fastest growing industries with growth rates over 40% per year [2]. In Germany, photovoltaics has overtaken the hydropower and supplied over 3% of the electricity consumption (almost 1% of the final energy consumption) in 2011 [3].

Among all photovoltaic technologies, silicon solar cells are still dominating the world market. However, thin-film solar cells are growing at an extremely rapid pace increasing their market share from 6% in 2005 [2] to 13% in 2010 [4]. Nowadays, the most effective way to reduce the price per watt, and thereby putting thin-film solar cells to a more competing position, is to increase the power conversion efficiency of the products. In all kinds of thin-film solar cells, $Cu(In,Ga)Se_2$ (CIGS) solar cells have shown the highest power conversion efficiency since more than 15 years [5]. The most recent world record of 20.3% was created by the "Zentrum für Sonnenenergie- und Wasserstoff-Forschung Baden-Württemberg" (ZSW) in 2011 [6]. This record is fully comparable to the best multicrystalline silicon solar cells (20.4% [7, 8]). Therefore, CIGS is generally considered to be the most promising material for low-cost productions [9].

In the last three decades, numerous measurement techniques were applied on CIGS solar cells, in order to find the key for their excellent performance. One of them is Kelvin probe force microscopy (KPFM). It enables a simultaneous measurement of topography and contact potential difference with spatial resolution in the nanometer range. The contact potential difference is directly correlated to the local electrostatic potential. As a solar cell converts radiative energy to electrical energy, the distribution of the electrostatic potential in and between the materials in the solar cell plays a crucial role in this conversion process. In the current study, the following four topics related to the distribution of the electrostatic potential are targeted by means of KPFM:

1. The highest power conversion efficiencies of the solar cells were achieved at [Ga]/([Ga]+[In])-ratios around 0.3 [6, 10] corresponding to band gap energies of ca. 1.2 eV. This value mismatches the estimated optimum value in the range of 1.4–1.5 eV [9]. By further increasing the Ga content the solar cell performance obviously degrades, mainly due to the limited open circuit voltage. Physically, the open circuit voltage is correlated to the diffusion voltage, which is the potential drop through the heterojunction of the solar cell. Therefore, special attention will be paid to the potential drop through the heterojunction in dependence of the Ga ratio.

2. Grain boundaries are usually regarded as detrimental for the performance of semiconductor devices [11]. However, despite of the abundance of grain boundaries, CIGS solar cells based on polycrystalline materials outperform their monocrystalline counterparts [12]. This interesting effect has been explained by various grain boundary models. Some models claimed that the potential variations at grain boundaries are beneficial for the collection and transport of the photogenerated charge carriers [13, 14]. In the current study, the

functionality of grain boundaries will be reevaluated based on KPFM measurements in darkness, and more importantly, under white light illumination.

3. The potential distribution through the heterojunction is very helpful for understanding the working principle of CIGS solar cells. Unfortunately, most of the previous work was done on samples that are no more functional as solar cells due to the preparation processes. As a consequence, the conclusions drawn in this way might be invalid for real devices. Therefore, the focus will be laid on the potential distribution through the heterojunction in operating CIGS solar cells. Moreover, the KPFM measurements will be carried out under different operating conditions of the solar cell, including in darkness and at defined illumination intensities with white light.

4. In order to abandon the heavy metal Cd and reduce the absorption losses in the buffer layer, extensive research efforts were made to find alternatives for the conventional CdS buffer layer. One of the most successful candidates is ZnS. By exchanging the CdS/i-ZnO buffer system with ZnS/(Zn,Mg)O, high performance Cd-free CIGS solar cells were fabricated. Unfortunately, the gain in the short circuit current as the result of reduced absorption losses is mostly accompanied by a decrease in the open circuit voltage. Given the correlation between the open circuit voltage and the potential drop through the heterojunction, the difference in the potential distribution in solar cells based on these two different buffer systems will be carefully analyzed.

KPFM studies on CIGS solar cells have been reported previously. However, three issues should be carefully reconsidered. First, CIGS grain boundaries on the surface of CIGS absorbers were studied, where the chemical composition is known to be different to that of the bulk material. However, grain boundaries in the bulk, which are more decisive for the solar cell performance, were not studied. Second, potential distributions were investigated on polished cross sections of the solar cell. However, the mechanical polishing process will modify the original properties of the cross sections, which may lead to inaccurate conclusions. Third, except Ref. [15] none of the former work was established with illumination of the solar cell through the front contact, as one usually does for operating a solar cell. Obviously, the illumination has a great significance for understanding the true working principle of CIGS solar cells.

In order to overcome these issues, a method is developed in this study. This method enables KPFM measurements on cleaved cross sections of CIGS solar cells. More importantly, since the samples prepared by this method are fully functional, their properties can be investigated under conditions similar to the test standard for solar cells. To the best knowledge of the author, this is the first study where polycrystalline solar cells in operation are analyzed by KPFM.

Encompassing the four aforementioned topics this thesis is structured as follows:

Chapter 2 introduces the solar cell structure and the deposition processes for individual layers. In addition, some physical fundamentals of CIGS solar cells relevant for the discussions in the later chapters are presented.

Chapter 3 gives an introduction to the working principle of KPFM measurements and some fundamental aspects of semiconductor surfaces. Moreover, important former studies are summarized, in order to clarify the improvements and breakthroughs realized in this work.

Chapter 4 describes the experimental setup of KPFM and the optimization of its major parameters. Furthermore, the preparation procedure of cleaved cross sections is introduced.

Chapter 5 shows the investigation on CIGS solar cells with varying Ga contents combining KPFM with two further measurement methods. The

Fermi energy shifting in CIGS absorbers in dependence of the Ga content is studied, leading to the discussion about different magnitudes of charge carrier recombination in these solar cells.

Chapter 6 compares the potential variations at grain boundaries on the surface and on untreated cross sections of a $CuIn_{0.7}Ga_{0.3}Se_2$ absorber. The comparison between the values achieved from these two positions evokes a discussion about the reevaluation of the conclusions in some former studies.

Chapter 7 shows the measurements of the potential distribution through the solar cell heterojunction in darkness and under defined illumination intensities with white light. This chapter then focuses on potential variations at grain boundaries under illumination, in order to understand the functionality of grain boundaries in CIGS solar cells in operation.

Chapter 8 gives a comparison between CIGS solar cells fabricated with ZnS/(Zn,Mg)O and CdS/i-ZnO buffer systems. The potential distributions through the solar cell heterojunction formed with different buffer systems are analyzed, in order to understand the loss of the open circuit voltage in solar cells with the ZnS/(Zn,Mg)O buffer system.

Chapter 9 makes some recommendations on the optimization of CIGS-based and other kinds of thin-film solar cells, on the basis of the acquired knowledge from the previous chapters.

Chapter 10, the last chapter, concludes the major results of this study and gives an outlook for the pursuing work.

2 Cu(In,Ga)Se$_2$ thin-film solar cells

In this chapter the history and development of Cu(In,Ga)Se$_2$ thin-film solar cells are briefly reviewed. The state of the art including the solar cell structure and the deposition techniques for individual thin-film layers are shortly introduced. In addition, some physical fundamentals are presented, which provide a theoretical basis for the experimental findings in the upcoming chapters. Finally, two measurement methods are introduced, which are generally applied for determing significant photovoltaic parameters of solar cells.

2.1 Historical background

The research of solar cells based on Cu(In,Ga)Se$_2$ started in 1975, when Shay *et al.* [18] from the Bell laboratories evaporated 5–$10\,\mu$m CdS onto a CuInSe$_2$ single crystal. A power conversion efficiency of 12% was not achieved under a calibrated solar simulator like today, but roughly determined "on a clear day in New Jersey". Shortly after that, Kazmerzki *et al.* at the University of Maine [19] demonstrated the first thin-film CuInSe$_2$/CdS solar cells. In that work the absorber layers were evaporated by a two-source (CuInSe$_2$+Se) technique, in order to control the fabricated absorber to be p-type. Two sample designs, i.e., illumination through CuInSe$_2$ or CdS, were utilized and efficiencies of the devices in the 4–5% range were presented. Interestingly, each device was subjected to a "short bake" in vacuum before the measurements were taken. This was maybe the primary form of the nowadays in the research and industry widely spread annealing processes for fabricating highly efficient CIGS solar cells.

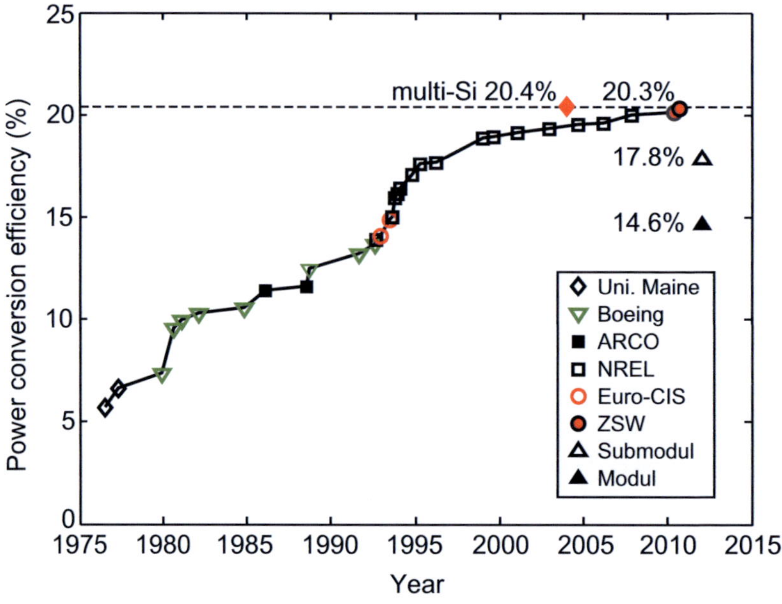

Figure 2.1: Record power conversion efficiencies of thin-film solar cells based on Cu(In,Ga)Se$_2$ (CIGS) demonstrated by various organizations. In the last thirty-five years the performance of this type of solar cell has steadily improved. The most recent record of the cell efficiency up to 20.3% (red circle) was demonstrated by ZSW, which shows hardly a difference to the best multicrystalline silicon solar cells (red diamond with dashed level line) [7,8]. The best CIGS submodule shows an aperture area efficiency of 17.8% [16] and the champion module has a total module efficiency of 14.6% [17]. The efficiencies of CIGS solar cells are extracted from Ref. [5].

Stimulated by the huge potential of CIGS solar cells exhibited in these primordial work, the industry jumped into this sector in the early 1980s. Two mainstream preparation concepts for the absorber layer were pursued, represented by Boeing and ARCO, respectively. The Boeing group developed a co-evaporation process where elemental copper (Cu), indium (In) and selenium (Se) were deposited from separate evaporation sources [20]. In contrast, ARCO used a two-step process. The first step was the subsequent deposition of three stacked precursor layers of Cu, In and Se. The Cu and In layers were prepared by means of electro-deposition and the Se layer on top by vacuum evaporation. The second step was the annealing of the precursor layers in a nitrogen gas atmosphere at elevated temperature [21]. Based on these two competing techniques, the record efficiencies were hold for almost 15 years by the industry (see Fig. 2.1). In the 1990s, a research alliance of 12 institutions in Europe, the EuroCIS [22], and the National Renewable Energy Laboratory (NREL) in the US, took over the relay baton for the record cell efficiency from the two industrial giants. Particularly, based on an advanced co-evaporation process, NREL has kept the record of cell efficiency for nearly 15 years. Since this process consists of three phases [23], it is called the three-stage process. This "monopoly" situation was terminated by Zentrum für Sonnenenergie- und Wasserstoff-Forschung Baden-Württemberg (ZSW) in 2010 by fabricating the first $Cu(In,Ga)Se_2$ solar cells with efficiencies beyond 20% [24]. The most recent record refreshed by ZSW in 2011 was up to 20.3% [6]. This encouraging result showed only negligible difference to the best multicrystalline silicon solar cells (20.4% [7, 8]). More importantly, the fact that CIGS solar cells on a 20% efficiency level can still be produced with varying composition pointed out the existing unexploited potential of this type of solar cell [6]. Around 2000 Würth Solar (Manz AG since 2011) in cooperation with ZSW brought the inline co-evaporation technology into mass production. Since then, a broad spectrum of companies worldwide has joined the competition. Nowadays, the most active module producers are Solar Frontier, AVANCIS, Q-Cells, Bosch CISTech, Miasolé, etc. After the technology transfer and condense in these years, the best submodule with an aperture area efficiency of 17.8% was demonstrated by Solar Frontier [16]. The champion module with a total module efficiency of 14.6% and an aperture area efficiency of 15.9% was recently fabricated by Manz AG [17].

2.2 Fabrication of Cu(In,Ga)Se$_2$ thin-film solar cells

2.2.1 Solar cell structure

Figure 2.2 is a false-color scanning electron microscopy (SEM) image of the cross section of a CIGS thin-film solar cell investigated in this work. This solar cell is fabricated at ZSW. In the fabrication a 3 mm thick soda-lime glass (invisible in the image) is utilized as the substrate. A molybdenum (Mo) layer with a thickness of 500 nm is deposited as the back contact by a sputtering process in argon (Ar) atmosphere. As the next step, a 2 μm thick CIGS absorber layer is deposited via a co-evaporation process. Subsequently, a buffer system consisting of a 50 nm thick cadmium sulfide (CdS) layer and a 50 nm thick intrinsic ZnO (i-ZnO) layer is deposited

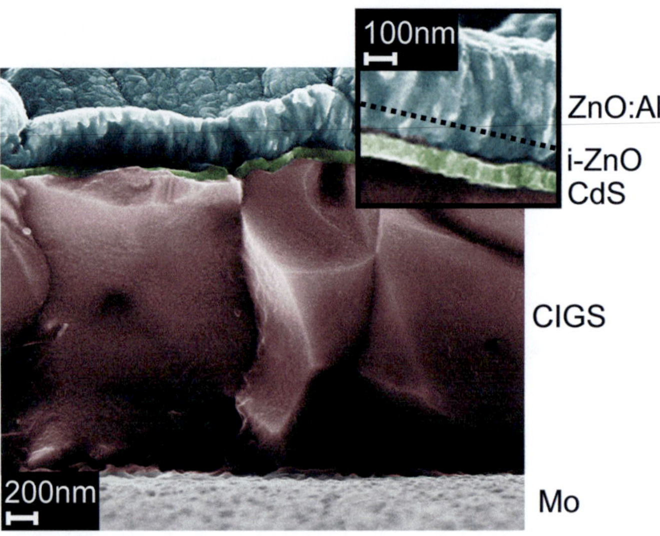

Figure 2.2: A false-color scanning electron microscopy image of the cross section of a CIGS thin-film solar cell investigated in this work. On a soda-lime glass substrate the Mo, CIGS, CdS, i-ZnO, ZnO:Al layers are subsequently deposited. For a good visibility, the heterojunction of the solar cell is zoomed in. The sample is fabricated at ZSW.

using chemical bath deposition (CBD) and radio frequency (rf) sputtering, respectively. Then, an aluminum doped ZnO (ZnO:Al) layer with a thickness of 350 nm is sputtered by means of dc-sputtering as the front contact. Finally, a finger structure of a nickel and aluminum alloy (Ni/Al) is evaporated with an electron gun, in order to improve the charge carrier collection and transport. This finger structure is not included in the SEM image. CIGS solar cells with a zinc sulfide/zinc magnesium oxide (ZnS/(Zn,Mg)O) alternative buffer system are also a subject of this work. In these samples the ZnS layer is similarly deposited by the CBD process. The (Zn,Mg)O layer is prepared by means of rf-sputtering and the [Mg]/([Mg]+[Zn])-ratio is 0.25 [25].

2.2.2 Deposition of Cu(In,Ga)Se$_2$ absorber layer

Benefiting from the uninterrupted work flow, dynamic inline deposition processes have the greatest importance for industrial production. CIGS solar cells based on absorbers fabricated in a highly optimized process at ZSW, the multistage inline co-evaporation process, have reached efficiencies as high as 19.6% [26]. The primary form of the inline process at ZSW was described in Ref. [27]. In that process, elemental Cu, In, Ga and Se were co-evaporated from separate line sources downwards onto $30 \times 30\,\mathrm{cm}^2$ Mo-coated glass substrates. The evaporation rates of individual elements were controlled by the atomic absorption spectroscopy (AAS) coupled to the deposition chamber with optical fibers. The substrates were heated at a fairly constant temperature and moved at a constant speed through the chamber. Different band gap gradients in the absorber layer could be obtained in this dynamic process by the spatial variation of the In and Ga elemental flux distributions along the substrate moving axis, i.e., by the design and geometric positioning of the line sources. Years later, this system was extended with an additional chamber, which allows individual adjustment of the material composition and the heater temperature at each deposition step [28]. Therefore, this process was named the multistage process. In this work, CIGS absorbers studied in Chap. 5 are fabricated in a single stage inline process that is highly similar to the process introduced in Ref. [27]. The substrate is kept at a nominal temperature of 550–570 °C. The Ga content is kept constant at defined ratios during the deposition. More details about this process was given by Witte *et al.* in Ref. [29]. All the other samples, i.e., the ones studied in Chap. 6, 7 and 8, are produced in the multistage process.

2.2.3 Deposition of CdS and ZnS buffer layers

The CdS and ZnS buffer layers in CIGS solar cells investigated in this work are fabricated by chemical bath deposition (CBD). For depositing CdS, cadmium sulfate ($CdSO_4$), ammonium hydroxide (NH_4OH), and thiourea ($SC(NH_2)_2$) are used as precursors. The composition of the chemical bath employed for ZnS is very similar solely exchanging $CdSO_4$ with zinc sulfate ($ZnSO_4$). Another difference for depositing ZnS is that the films are rinsed with NH_4OH after the CBD process, in order to remove the superfluous zinc hydroxide ($Zn(OH)_2$) from the layer surface [25]. As reported by Contreras *et al.*, the ZnS films prepared by CBD process contain amounts of oxygen in the form of $Zn(OH)_2$ and ZnO [30]. Hence, the layers are also formally described in some publications as ZnS(O,OH) or Zn(S,O,OH) [25, 30]. For simplicity, the notation ZnS is used in this study.

2.3 Physical properties of Cu(In,Ga)Se$_2$ thin-film solar cells

2.3.1 Energy band diagram of Cu(In,Ga)Se$_2$ thin-film solar cells

Figure 2.3 shows the qualitative energy band diagrams of a CIGS solar cell in darkness and under illumination. The solar cell has a conventional Mo/CIGS/CdS/i-ZnO/ZnO:Al/Al-Ni-alloy layer stacking and is electrically in open circuit condition. E_{vac}, E_C, E_F and E_V are the vacuum energy, the

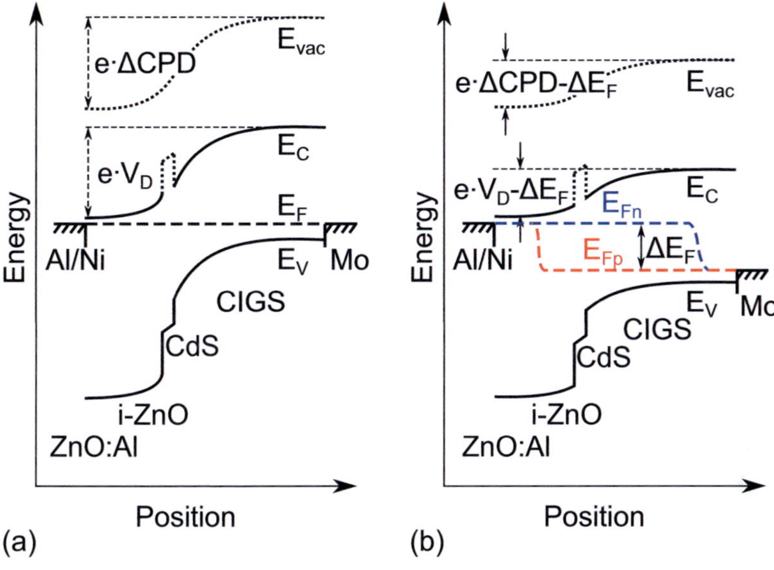

Figure 2.3: Qualitative energy band diagrams of a CIGS solar cell in open circuit condition. **(a)** In darkness the Fermi energies in different layers are evened out at the same level E_F. Band bending is formed in the vacuum energy E_{vac}, the conduction band energy E_C and the valence band energy E_V. ΔCPD and the diffusion voltage V_D describe the potential drop in E_{vac} and E_C, respectively. **(b)** Under illumination the Fermi energy splits into the quasi-Fermi levels E_{Fn} and E_{Fp}. Band bending in the energy bands E_{vac}, E_C and E_V is reduced by $\Delta E_F = E_{Fn} - E_{Fp}$.

conduction band energy, the Fermi energy and the valence band energy, respectively. V_D is the diffusion voltage, which is by definition the maximal potential drop in the conduction band E_C in darkness and without external bias [31]. As V_D is the theoretical upper limit of the open circuit voltage [32], it has a great importance to the solar cell performance. ΔCPD is the difference between the CPD (contact potential difference) values of ZnO and CIGS. As it will be discussed in more detail in Chap. 5, ΔCPD is proportional to the work function difference between ZnO and CIGS by a factor of the elemental charge e. In the band diagrams ΔCPD describes the potential drop in the vacuum level E_{vac}.

In darkness the solar cell is in equilibrium and the electrochemical energy is overall constant. Thus, the Fermi energies in different materials are located at the same level E_F and band bending is formed in the energy bands E_{vac}, E_C and E_V. If the solar cell is illuminated with photons with energies larger than the band gap energy, electron-hole pairs will be generated and separated in the CIGS absorber. Driven by the concentration gradient, the photogenerated electrons will diffuse into the space charge region (SCR) and will be transferred due to the electrical field in SCR to the ZnO layer, whereas the photogenerated holes accumulate in the CIGS layer. Due to the redistribution of the free charge carriers, the Fermi energy E_F splits into two quasi-Fermi levels E_{Fn} and E_{Fp}, one each for the electrons and holes. Their splitting results in a photovoltage between the edges of the CIGS/CdS/ZnO-heterojunction, which is equal to $\Delta E_F/e = (E_{Fn} - E_{Fp})/e$. This photovoltage can be practically measured between the electrodes of the solar cell. Meanwhile, band bending in the energy bands E_{vac}, E_C and E_V is flattened by ΔE_F. Accordingly, the potential drop in the energy bands is reduced by $\Delta E_F/e$ [33].

2.3.2 Ga content in Cu(In,Ga)Se$_2$ absorber layer

The systematic investigation of the impact of the Ga addition on CuInSe$_2$ started in the middle of 1980s [34]. It was found that the band gap energy of the CIGS absorber increases almost linearly with the [Ga]/([Ga]+[In])-ratio (GGI). Upon this important finding, the efficiency of the solar cell could be noticeably increased with proper Ga addition due to a closer match to the optimum value for absorbing the terrestrial solar spectrum [34–36]. Up to date, the most efficient CIGS solar cells were fabricated with GGI ratios around 0.3 [6, 10]. At larger Ga contents it was reported that the open circuit voltage V_{oc} does not increase proportionally to the band gap energy and consequently the efficiency significantly degrades [37, 38]. Indeed, CIGS solar cells with higher Ga contents may bring the following advantages. First, the band gap energy will further approach the optimum value of 1.4–1.5 eV [9], which promises an even higher device efficiency. Second, solar cells with higher Ga contents will generally deliver higher voltages and smaller currents, and as a result lower resistive losses in operation, which is crucial for highly efficient modules and systems. Third, absorbers with a higher Ga content are good candidates for the top cell in tandem solar cells exclusively based on CIGS absorbers. Consequently, extensive research efforts have been devoted on CIGS solar cells with GGI ratios beyond 0.3. Particularly, the defect characteristics and the resulting recombination processes have been widely investigated in a huge number of publications [39–59]. It has been commonly accepted that the unsatisfying performance of CIGS solar cells with high Ga content is primarily limited by high recombination rates of the free charge carriers. However, the dominating recombination process, i.e., the bulk recombination in CIGS absorber [42, 45, 46, 48, 49] or the interface recombination at the buffer/CIGS-interface [39, 47, 51, 56–58], is still a matter of debate. The dominating process seems to depend strongly on the quality of the CIGS absorber and the buffer layer.

2.3.3 Defect characteristics of Cu(In,Ga)Se$_2$ material

Cu(In,Ga)Se$_2$ compounds are well known for their capability to tolerate large deviations from stoichiometry without impediment of their photovoltaic properties. This makes it possible to use relatively simple technology for preparation of the absorber layer without strict control over its composition. The compositional variations from stoichiometry are directly related to the formation of native defects. In a ternary system like CuInSe$_2$, already 12 native defects are conceivable: three vacancies, three interstitials and six antisite defects. Moreover, any complex of these defects could occur [48]. With the addition of Ga, the amount of possible defects is again noticeably increased. Indeed, the situation in a real Cu(In,Ga)Se$_2$ solar cell is even more complicated, since foreign elements that outdiffuse from other layers like cadmium (Cd), magnesium (Mg), zinc (Zn), sodium (Na) and oxygen (O) may all contribute to the formation of additional defect levels [60–62]. Among them there are acceptor and donor defects. Their concentrations depend on material composition and defect formation enthalpies [63]. The net doping of the material results from the difference in the concentrations of acceptor and donor defects. It determines the diffusion voltage of the solar cell and therefore limits the open circuit voltage and the width of the space charge region [62].

Physically, by increasing the doping density the free carrier concentration will increase. However, there is a maximum for the doping level in each given semiconductor. This maximum does not depend on the dopant species or the method by which the dopants are introduced into the semiconductor. It is an intrinsic property of the material [48]. This effect is explained as the self-compensation in semiconductors. It describes that at high doping levels the dopants begin to occupy both acceptor and donor sites that compensate each other. The self-compensation effect was also studied in Cu(In,Ga)Se$_2$ materials. The analysis of the temperature-dependent charge carrier concentration showed that the ratio between the concentrations of acceptors and donors approaches one with increasing acceptor concentration [64]. It was argued that when more acceptors are created, the Fermi level shifts down and the formation enthalpy of the donors decreases. This in turn will create more donors. The research on the self-compensation effect in Cu(In,Ga)Se$_2$ materials grown under different compositions showed that Cu rich materials are already compensated and the degree of compensation is even much higher in Cu poor materials [65,66], which are widely used as absorbers in highly efficient solar cells.

2.3.4 Cu(In,Ga)Se$_2$ grain boundaries

Interest in grain boundaries (GB) in semiconductors has undergone a remarkable expansion in the last decades. The reason for this can be traced to the development of polycrystalline semiconductors as convenient low-cost candidates for photovoltaics and other applications. A GB is the interface between two grains in a polycrystalline material with a certain density of structural defects. A common classification scheme for GBs is the coincidence site lattice notation [67]. The coincidence site lattice is constructed as follows: at a GB two lattices of different orientations join. Virtually, one lattice can be considered to extend into the volume space of the other one. In the virtually common volume of the two lattices there will be lattice sites which coincide. These coincident sites will again form a periodic structure, which has a larger elementary cell than that of the original crystal lattice. The volume ratio between the coincidence site lattice and the original crystal lattice gives the Σ value of the GB. In general, a larger Σ value indicates a larger density of broken and distorted bonds and vice versa [63].

As structural defects in ideally grown semiconductors, GBs are normally regarded as introducing deleterious effects such as lower electrical and thermal conductivity and higher recombination rates of the free charge carriers. However, GBs in Cu(In,Ga)Se$_2$ materials appear to be exceptions, since solar cells fabricated from polycrystalline CIGS absorbers outperform their monocrystalline counterparts [12]. In order to disclose possible beneficial effects of GBs in CIGS absorbers, various characterization techniques have been applied on CIGS absorbers or completed solar cell devices. These techniques include transmission electron microscopy [68], micro-Auger spectroscopy [69], cathodoluminescence spectroscopy [69–71], electron backscatter diffraction [71], atomic probe tomography [72], and Kelvin probe force microscopy, which is the key method of this work. Due to the small dimension of GBs in CIGS absorbers, all of these feasible techniques have to possess spatial resolutions in nanometer range.

Based on results from these measurements, different electrical models for GBs have been built, which were comprehensively reviewed in Refs. [12, 63]. Due to different technical limitations of the measurement techniques and different samples under investigation, the conclusions based on different models were usually inconsistent to each other. Nevertheless, after the last few years several renowned research groups slowly approached a consensus that most GBs in CIGS absorbers are rather inactive for free carrier recombination [12, 73–76], which already provides a good base for reasonable photovoltaic devices [12].

2.4 Measurement of the efficiency of solar cells

2.4.1 Current density-voltage measurement

Extensive efforts were made to improve the performance of solar cells. Indeed, this so-called performance can be defined in multiple ways. The most important one is the power conversion efficiency η, which is by definition the ratio of the maximum electrical power density P_{el}^{max} to the incident optical power density P_{opt}. Practically, P_{el}^{max} is rated in the current density-voltage (jV) measurement under the standard test conditions (STC), i.e., irradiance of $P_{opt} = 100\,mW/cm^2$, spectral irradiance of air mass 1.5 (AM 1.5) and cell temperature $T = 25\,^\circ C$.

Figure 2.4 depicts the jV-curves of a solar cell in darkness and under illumination. The short circuit current density j_{sc} is the current flow normalized by the solar cell area in short circuit condition (V=0). The open circuit voltage V_{oc} is the voltage between the electrodes in open circuit condition (j=0). Obviously, in these two points the electrical power density defined as $P_{el} = j \cdot V$ is zero. Between these two points there is a maximum power point (MPP), where the solar cell delivers the maximum power density P_{el}^{max}. Based on the quantities mentioned above, the fill factor FF can

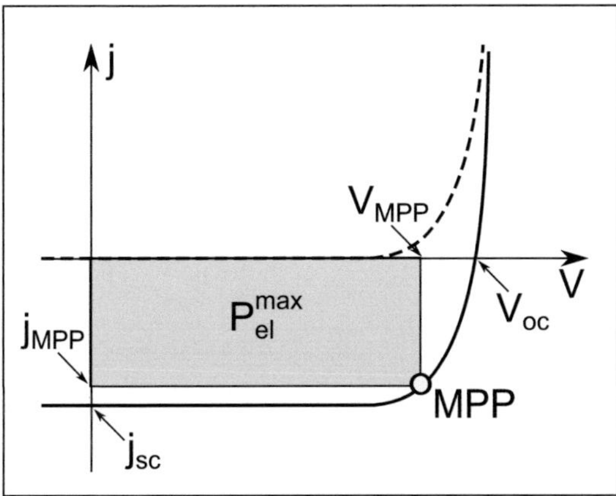

Figure 2.4: Current density-voltage (jV) curves of a solar cell in darkness (dashed line) and under illumination (solid line). j_{sc} and V_{oc} are the short circuit current density and the open circuit voltage, respectively. In the maximum power point MPP the solar cell delivers the maximal power density P_{el}^{max}.

be described as:

$$FF = \frac{P_{el}^{max}}{j_{sc} \cdot V_{oc}} = \frac{j_{MPP} \cdot V_{MPP}}{j_{sc} \cdot V_{oc}} \qquad (2.1)$$

Ultimately, the power conversion efficiency η is defined as:

$$\eta = \frac{P_{el}^{max}}{P_{opt}} = \frac{FF \cdot j_{sc} \cdot V_{oc}}{P_{opt}} \qquad (2.2)$$

2.4.2 External quantum efficiency measurement

Another important efficiency for solar cells is the external quantum efficiency (EQE). The EQE measurement is an extremely useful tool for the identification of different loss mechanisms in solar cells, e.g., the surface recombination, the bulk recombination due to low diffusion length and the back contact recombination. It is defined as the ratio of the number of charge carriers collected by the solar cell to the number of incident photons of a given wavelength, or that of the photo current (charge carriers per second) to the photon flux (photons per second). In this work EQE measurements were used to determine the optical band gap energy E_g of the CIGS absorbers. Physically, EQE at a given photo energy $h\nu$ can be expressed as [77]

$$EQE\,(h\nu) = 1 - \frac{\exp\left(-\alpha W\right)}{\alpha L_{eff} + 1} \tag{2.3}$$

with the absorption coefficient α at the photo energy $h\nu$, the depletion width W of the solar cell and the effective diffusion length L_{eff} for minority charge carriers. For photo energies near the band gap energy, the term αL_{eff} is smaller than 1 (based on $\alpha < 10^{-4}\,\mathrm{cm}^{-1}$ for CuInSe$_2$ [78] and $L_{eff} \approx 1\,\mu\mathrm{m}$ [79]). Consequently, Eq. 2.3 can be reduced to

$$EQE\,(h\nu) = 1 - \exp\left(-\alpha W\right) \tag{2.4}$$

and thus

$$\alpha = \frac{1}{W}\ln\left(1 - EQE\,(h\nu)\right) \tag{2.5}$$

For a direct transition, the dependence of the absorption coefficient on the photo energy is given by [80]

$$\alpha h\nu \propto \left(h\nu - E_g\right)^{1/2} \tag{2.6}$$

By substituting α in Eq. 2.6 with Eq. 2.5 and taking the square function of both sides, the following relation is available

$$\left[h\nu \times \ln\left(1 - EQE\,(h\nu)\right)\right]^2 \propto h\nu - E_g \tag{2.7}$$

Therefore, the band gap energy E_g can be extrapolated with the expression $\left[h\nu \times \ln\left(1 - EQE\,(h\nu)\right)\right]^2$, which is a function of the photo energy $h\nu$.

3 Kelvin probe force microscopy

Kelvin probe force microscopy (KPFM) is a scanning probe microscopy technique, which enables a simultaneous measurement of topography and contact potential difference (CPD). By applying an extremely sharp tip as a sensor, KPFM can achieve spatial resolution in the nanometer range. Therefore, KPFM is a powerful tool for studying the electrical properties in CIGS thin-film solar cells, which consist of layers as thin as only few micrometers. This chapter starts with a review of the history of this technique. Subsequently, the working principle of KPFM and some fundamental aspects of semiconductor surfaces are introduced. Finally, important previous studies are summarized, in order to clarify the improvements and breakthroughs in the current work.

3.1 Historical background

Surface science has experienced a revolution by the invention of the scanning tunneling microscope (STM) in 1982 [81]. Consequently, its inventors Gerd Binnig and Heinrich Rohrer were awarded with the Nobel prize in 1986. With this technique the first images showing atomic resolution on a Si(111)7x7 surface were obtained. However, due to its working principle – detecting the tunneling current between the tip and sample surface under an external voltage – the STM is inherently limited to the study of conducting surfaces. Four years later, the invention of the atomic force microscope (AFM) by Binnig *et al.* overcame this limitation and widened the measurable samples to non-conductive ones [82]. This method works by measuring the static deflection of the tip supported by a cantilever beam and is thus called contact mode AFM. Shortly after that, Martin *et al.* [83] improved the resolution by vibrating the cantilever at a frequency near its resonance frequency. They noticed that the amplitude of the tip's oscillation changed with the tip-sample distance. Consequently, by using a feedback loop, which keeps the oscillation amplitude constant through adjusting the tip-sample distance, the topography could be recorded. A direct contact of the tip and sample can be avoided in this way and the forces between them can be effectively minimized. Therefore, this mode was named non-contact AFM and was widely employed for measurements on soft samples, e.g., biological and polymer samples.

Later on, the combination of AFM with other measurement methods has provided a wide range of possibilities to access additional sample properties with nanometer resolution. The most successful ones include magnetic force microscopy (MFM) [84], electrostatic force microscopy (EFM) [85] and scanning capacitance microscopy (SCM) [86]. The method applied in this study is Kelvin probe force microscopy (KPFM), which was developed by Nonnenmacher *et al.* in 1991 [87]. It allows the simultaneous measurement of topography and contact potential difference (CPD) between tip and sample. Basically, this method can be interpreted as a miniature of the macroscopic Kelvin probe method that was named after its inventor Lord Kelvin [88]. In his original experiments in 1898, a capacitor consisting of two parallel plates from different metals was connected by an electrometer. It was found that a displacement of either metal plates will cause a charge flow into the electrometer. Furthermore, by applying a dc-voltage between the metal plates and adjusting its value until the displacement of the metal plates produces no charge flow any more, the CPD between the two metals could be determined. At that time, for each varied value of the dc-voltage, the electrometer had to be discharged and therefore single CPD determination even with limited precision required several minutes. In 1932, the introduction of a vibrating capacitor by Zisman *et al.* significantly speeded up the measurement velocity [89]. In his setup, one metal plate of the capacitor was vibrated periodically and thus an ac current was generated, which could be continuously monitored. The dc-bias could be easily adjusted until the ac current was nullified. However, the reduction of this principle to the microscopic scale would result in a poor sensitivity, since the capacitor formed by the tip and sample surface is too small to generate a sufficient current. Therefore, as it will be introduced in more detail in Chap. 3.2.3, in the modern KPFM setup the electrostatic force between the tip and sample is detected providing a much higher measurement sensitivity.

3.2 Measurement principle

3.2.1 Topography signal acquired by tapping mode atomic force microscopy

In AFM a sharp tip at the end of a cantilever serves as a force sensor. The topography of the sample is imaged by scanning the tip over the sample surface while a constant force or force gradient is maintained by a feedback loop. When the tip approaches a surface, it "feels" a mixture of different forces consisting of the short-range chemical force, the van der Waals force, and the long-range electrostatic force and magnetic force. For KPFM, the most relevant forces are the van der Waals force and the electrostatic force. In the following, the common modes employed in AFM will be briefly introduced. Subsequently, the tapping mode feasible for KPFM will be shown in more detail.

It was pointed out that the ensemble of AFM suffers from a problem of terminology [90]. This means that there are different ways to categorize all the working modes. In this work the working modes are categorized depending on the tip-sample distance and the oscillation amplitude of the cantilever into the contact mode, non-contact mode and tapping mode as sketched in Fig.3.1.

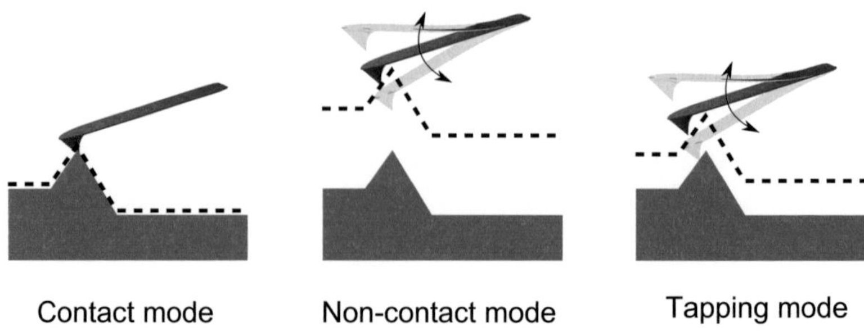

Figure 3.1: Depending on the tip-sample distance and the oscillation amplitude of the cantilever, atomic force microscopy can be performed in the contact mode, non-contact mode and tapping mode. In contact mode the cantilever basically scratches over the sample surface and its static deflection is recorded for the topography profiling. In non-contact mode and tapping mode the cantilever is vibrated mechanically near its free oscillation frequency. Both modes are physically very similar except that in tapping mode the cantilever makes intermittent contact with the sample surface. The dashed lines in the sketches indicate the averaged position of the tip over the sample surface in these three modes.

Contact mode

The contact mode is based on the measurement of the static deflection of the cantilever. Topography images are recorded by scanning the tip over the sample surface at a constant cantilever deflection. In this mode, the tip is in direct contact with the sample surface, i.e., physically in the repulsive force regime. Therefore, the contact mode usually gives a very high spatial resolution. However, the direct contact has the disadvantage that the surface of soft samples could be easily damaged. Nowadays, this mode has found further applications besides the measurements of topography, e.g., measurement of the friction with atomic resolution through the torsional bending of the cantilever [91] or nanolithography [92].

Non-contact mode

In the non-contact mode the cantilever is oscillated at or near its eigenfrequency. This oscillation is mechanically excited using a piezoelectric element (dither piezo or shaker) on which the cantilever-chip is mounted. In this mode the tip has no direct contact with the sample surface and is physically in the attractive regime of the tip-sample interaction. As a result, the forces exerted by the tip on the sample is considerably reduced in comparison with contact mode, which is highly beneficial for soft samples. Since the non-contact mode is in principle quite similar to the tapping mode, the main physical processes are introduced for the tapping mode in the following.

Tapping mode

In the tapping mode the cantilever is also mechanically oscillated at its eigenfrequency. Differently to the non-contact mode, the tip experiences an intermittent contact with the sample surface during each cycle and thereafter the nomination of this mode. The eigenfrequency, or the free resonance frequency f_0, is a function of the spring constant k and the effective mass m^* [93]:

$$\omega_0 = 2\pi f_0 = \sqrt{\frac{k}{m^*}} \tag{3.1}$$

As previously mentioned, if the cantilever is brought to the proximity of the sample surface, various forces will be exerted on it. Among them the van der Waals force and the electrostatic force contribute mostly to the KPFM studies.

Van der Waals forces are dipole-dipole forces. These forces are always present between atoms and molecules and arise from electrostatic dipoles induced by electromagnetic field fluctuations. The van der Waals forces between macroscopic bodies can be calculated in different ways. In the case of AFM the situation can be well approximated as a sphere, representing the tip, approaching an infinite plane, representing the sample surface. The van der Waals force can be then expressed as [94]:

$$F_{vdW} = -\frac{HR}{6d^2} \tag{3.2}$$

where H denotes the Hamaker constant, R the tip radius and d the closest distance between the sphere and the plane. Materials with high dielectric constant, e.g., metals have much larger H and as a result larger F_{vdW}. Moreover, the medium between the tip and sample has a great influence on F_{vdW} [94].

Electrostatic forces act between localized charges and distance dependently obey Coulomb's law. For conductive tips and conductive samples if the tip-sample configuration is considered as a plane capacitor and the electrical energy stored in the capacitor is $E_{el} = \frac{1}{2}CV^2$, the attractive force induced by the static charges can be derived as:

$$F_{el} = -\nabla E_{el} = -\frac{1}{2}\frac{\partial C}{\partial z}V^2 - CV\frac{\partial V}{\partial z} \tag{3.3}$$

with C the capacitance and V the electrical potential. Since V does not change with z, equation 3.3 can be simplified as [93]:

$$F_{el} = -\frac{1}{2}\frac{\partial C}{\partial z}V^2 \tag{3.4}$$

Now, if the tip approaches the sample surface, the interaction forces will cause a shift of the resonance curve of the cantilever. For small oscillation amplitude the shift of the resonance curve can be approximated by introducing an effective spring constant k_{eff} [95]:

$$k_{eff} = k - \frac{\partial F_{ts}}{\partial z} \tag{3.5}$$

where F_{ts} stands for the tip-sample interaction that originates from all exerted forces. For small force gradients the shift of the resonance frequency

Δf_0 can be easily derived from Eq. 3.1 and 3.5 as [95, 96]:

$$\Delta f_0 = -\frac{f_0}{2k}\frac{\partial F_{ts}}{\partial z} \tag{3.6}$$

This relation shows that the frequency shift is proportional to the force gradient.

In the tapping mode the cantilever is excited at its eigenfrequency. During the measurement, the change of the tip-sample distance leads to a change of the force gradient, which according to Eq. 3.6 results in a change of the resonance peak. Consequently, the oscillation amplitude at the fixed driving frequency changes. Therefore, the topography image can be achieved by keeping the oscillation amplitude constant through adjusting the tip-sample distance. As the measurement is based on the amplitude modulation (AM), the tapping mode is also called AM mode. Indeed, KPFM can be also performed through frequency modulation (FM), namely in FM mode. In FM mode the oscillation amplitude of the cantilever is maintained constant through an automatic gain control circuit, while the resonance frequency is measured directly by a frequency demodulator. In this way the tip-sample distance is adjusted during the measurement to keep a constant frequency shift Δf_0 relative to the free oscillation frequency. Physically, the response time of the system is in dependence with the oscillation damping of the cantilever and can be expressed as $\tau = 2Q/\omega_0$ [96]. Consequently, the AM mode is usually performed in air, where the quality factor Q is on the order of 100, whereas in vacuum, particularly in ultra-high vacuum (UHV), the scanning speed is noticeably slowed down due to very high values for Q factors typically above 10^5. Therefore, for applications under vacuum the FM mode is usually applied. In the current work the experimental setup is located in ambient conditions. Hence, all measurements were performed in AM mode.

3.2.2 Contact potential difference determined by macroscopic Kelvin probe method

Although the technical realization for detecting the tip-sample interaction in modern KPFM differs a lot from that Lord Kelvin performed between two metal blocks over 110 years ago, the physical principle remains the same. The work function ϕ is by definition the minimum energy required to remove an electron from the interior of a solid to a position outside of the solid surface [97]. In the energy diagram it is the energy difference between the vacuum energy E_{vac} and the Fermi energy E_F [98]. In Fig. 3.2 (a), (b) two materials with different work functions ϕ_1 and ϕ_2 are brought into contact. The electrons from the material 2 with a smaller work function will flow to the energetically more favorable position, namely material 1 with a higher work function, until the Fermi energies E_F in both materials are evened out. A contact potential difference (CPD) $V_{CPD} = (\phi_1 - \phi_2)/e$ between the local vacuum levels arises due to opposite charges with equal quantities on both sides. If an external voltage V_{dc} is applied between the materials, the accumulated charges and as a result the potential difference will be compensated. In the case of $V_{dc} = V_{CPD}$ the electrical field is exactly nullified. Thus, the work function of one material, e.g., ϕ_1 can be derived with V_{dc} and a known ϕ_2 of the other material.

3.2.3 Single mode Kelvin probe force microscopy

In modern KPFM, tips with radii of few nanometers are employed as sensors to enable a high spatial resolution. Despite of the extreme scale-down compared to the metal plates, the tip-sample system can still be well approximated as a tiny capacitor with the electrical capacitance of C. The electrostatic force between the electrodes of the capacitor was derived in Eq. 3.4. In KPFM an ac-voltage is applied between the tip and sample to modulate the electrostatic force. This ac-voltage has an amplitude of V_{ac} and an angular frequency of ω_{ac}. Additionally, a dc-voltage V_{dc} is applied

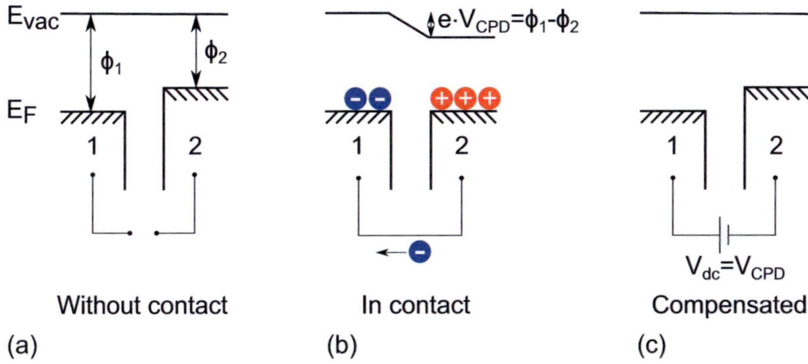

Figure 3.2: The energy band alignments of two materials with different work functions ϕ_1 and ϕ_2 in three situations: **(a)** without contact, **(b)** in contact and **(c)** with an external bias $V_{dc} = (\phi_1 - \phi_2)/e$.

to compensate the contact potential difference V_{CPD}. Therefore, Eq. 3.4 is now extended to

$$F_{el} = -\frac{1}{2}\frac{\partial C}{\partial z}\left[V_{dc} - V_{CPD} + V_{ac}sin(\omega_{ac}t)\right]^2 \qquad (3.7)$$

The product in Eq. 3.7 can be separated into three terms $F_{el} = F_{dc} + F_{\omega_{ac}} + F_{2\omega_{ac}}$ with

$$F_{dc} = -\frac{\partial C}{\partial z}\left[\frac{1}{2}\left(V_{dc} - V_{CPD}\right)^2 + \frac{V_{ac}^2}{4}\right], \qquad (3.8)$$

$$F_{\omega_{ac}} = -\frac{\partial C}{\partial z}\left(V_{dc} - V_{CPD}\right)V_{ac}\sin(\omega_{ac}t), \qquad (3.9)$$

$$F_{2\omega_{ac}} = \frac{\partial C}{\partial z}\frac{\partial V_{ac}^2}{\partial 4}\cos\left(2\omega_{ac}t\right). \qquad (3.10)$$

As it can be easily observed, F_{dc} is a static term, while $F_{\omega_{ac}}$ and $F_{2\omega_{ac}}$ oscillate with ω_{ac} and $2\omega_{ac}$, respectively. F_{dc} gives a constant additional deflection of the cantilever and $F_{2\omega_{ac}}$ can be used for capacitance microscopy. The contact potential difference V_{CPD} can be determined by minimizing $F_{\omega_{ac}}$. Practically, there are two methods to carry out KPFM. One method is to acquire the topography and CPD signals simultaneously in one single scan and is thus called the single mode. With the other method two scans are necessary. In the first scan the topography signal is acquired. In the second scan the cantilever is lifted up by several tens of nanometers and the CPD signal is acquired during retracing the saved topography signal. Due to the lifting of the cantilever, this method is named the lift mode. With the experimental setup in the present work, both modes can be performed. At the beginning of the setup optimization, both modes were compared [99]. It was figured out that the resolution of the CPD images is much better in the single mode. In a very recent publication Sadewasser *et al.* has summarized some major disadvantages of the lift mode [93]. First, the piezo creep or thermal drift makes it impossible to scan over exactly the same positions in the second pass. Second, the image resolution suffers from the increased tip-sample distance, because the tip lifted up from the sample surface averages over a larger area. Third, the local electrostatic forces are not compensated in the first pass, which in accordance with Eq. 3.8 leads to different additional deflection of the cantilever and therefore errors in the topography signal. Consequently, the single mode KPFM is employed throughout the current study.

In single mode KPFM, the cantilever is mechanically vibrated at its free resonance frequency f_0 by applying an ac-voltage $V_{ac}(f_0)$ on the shaker mounted at the end of the cantilever. Additionally, a second ac-voltage $V_{ac}(f_1)$ with a frequency f_1, approximately 6 times of f_0 [100], and a dc-voltage V_{dc} are applied on the sample surface. The cantilever is thus modulated with f_0 and f_1. As a result, the time signal of its oscillation amplitude captured by the 4-Quadrant detector is a mixture of two sine functions as displayed in Fig. 3.3 (a). Correspondingly, the power spectral density which is the fast Fourier transform of the time signal in Fig. 3.3 (b) shows two maxima at f_0 and f_1, respectively. With two lock-in amplifies the oscil-

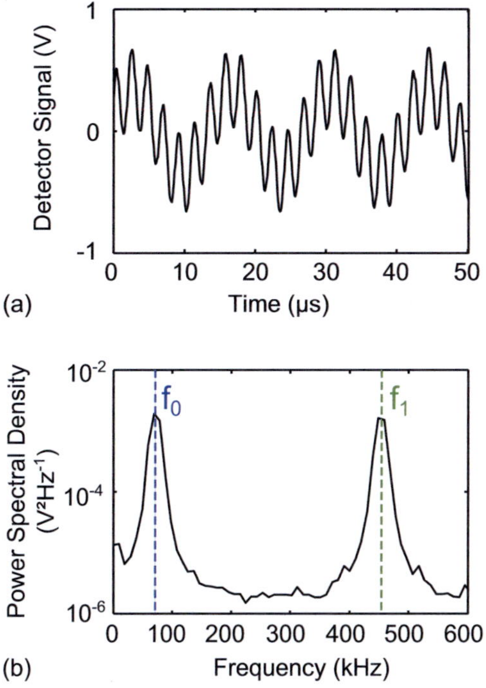

Figure 3.3: In single mode KPFM the cantilever is vibrated mechanically at its free resonance frequency f_0 and electrically at the second resonance frequency f_1. In addition a dc-voltage V_{dc} is applied between the tip and sample. Consequently, **(a)** the time signal of the oscillation amplitude of the cantilever is a mixture of two sine functions. Correspondingly, **(b)** the power spectral density which is the fast Fourier transform of the time signal shows two maxima at f_0 and f_1, respectively. With two lock-in amplifiers the oscillation amplitude A_0 and A_1 at f_0 and f_1 can be separately extracted. During the measurement, A_0 is kept constant by adjusting the tip-sample distance to achieve the topography profiling, while A_1 is minimized by adjusting V_{dc} to acquire the CPD signal.

lation amplitude of the cantilever A_0 and A_1 at f_0 and f_1 can be separately extracted. As introduced in many literatures (e.g. Ref. [101]), the lock-in amplifier or sometimes referred to as the phase sensitive detector is a measuring instrument that effectively responds to signals which are coherent (the same frequency and phase) with the reference signal and rejects all others. In other words, it extracts and amplifies weak signals from a noisy background. During the KPFM measurement, A_0 is kept constant by adjusting the tip-sample distance to achieve the topography profiling, while A_1 is minimized by adjusting V_{dc} to acquire the CPD signal.

3.3 Limiting factors for potential contrast

3.3.1 Energy band modification on clean semiconductor surfaces

Surface states induced surface space charge region

The periodic structure of the atoms in an ideal semiconductor results in the allowed energy bands, namely the conduction band E_C and the valence band E_V. Between them there is an energy gap E_g, where no electron states are allowed [102]. In the bulk of an ideal semiconductor, the probability to find an electron in any unit cell is equal because of the perfect three-dimensional translational symmetry. However, at the surface this symmetry is terminated in the direction perpendicular to the surface. Consequently, the unit cells in the vicinity of the surface are in general not identical to those in the bulk and additional surface-localized states may arise. In Fig. 3.4 (a) the situation for a p-type semiconductor with surface donor states is illustrated. Additional and more complex phenomena may contribute to the formation of surface-localized states. These phenomena include: dangling bonds, i.e., non-saturated chemical bonds of the surface atoms; surface reconstruction and relaxation, i.e., rearrangement of the position and/or chemical bonding configuration of surface atoms that minimizes the surface energy; impurity atoms adsorbed on the surface [103]. Surface-localized states caused by all the aforementioned effects induce the charge transfer between the bulk and surface until the thermal equilibrium is reached. This charge transfer results in a non-neutral region extended from the surface into the bulk, which is usually referred to as the surface space charge region (SSCR) in litera-

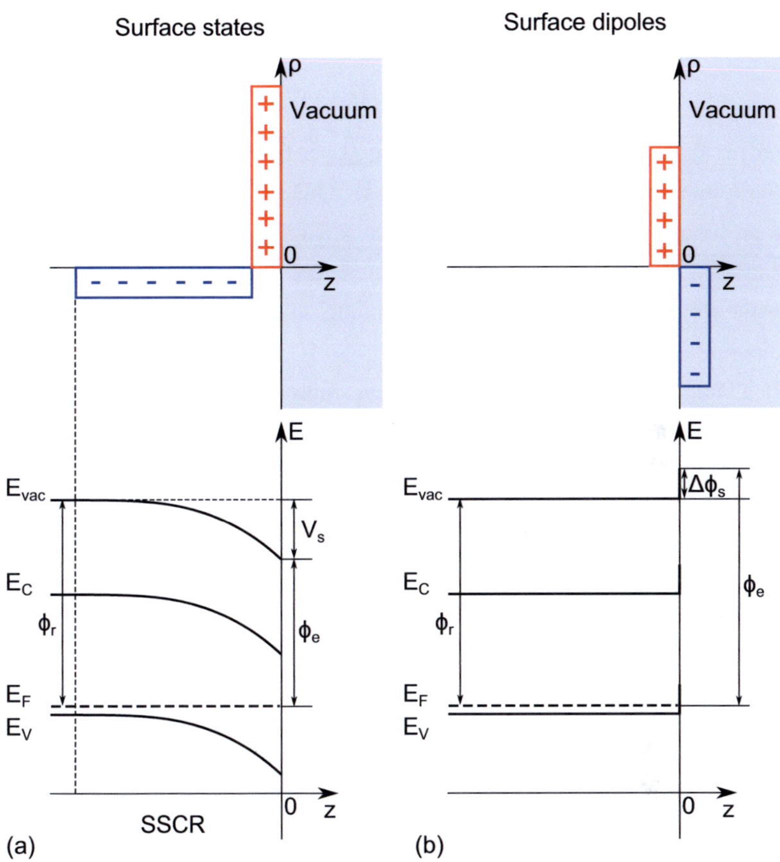

Figure 3.4: **(a)** Surface energy band bending induced by surface states. A non-neutral region is formed in the bulk near the surface, which is referred to as the surface space charge region (SSCR) [103]. **(b)** Surface band offset induced by surface dipoles. Due to the existence of the surface states and dipoles, the work function ϕ_e on the surface is usually different to ϕ_r, which stands for the energy difference between the Fermi energy E_F in the bulk of the material and the vacuum energy E_{vac} far away from the surface.

ture [103]. Obeying the charge neutrality rule, the following condition is reached under equilibrium:

$$Q_{ss} = -Q_{sc} \tag{3.11}$$

where Q_{ss} is the net surface charge and Q_{sc} is the net charge in the SSCR (both measured as charges per unit area). Quite similar to the situation at a pn-junction, the electrical field in the SSCR causes a gradual potential variation over this region with a total potential drop of V_s. From an energetic point of view, a surface band bending will be induced in the energy diagram as shown in Fig. 3.4 (a).

In a simplified situation, where only one single state with the energy level E_t exists, the correlation between Q_{ss} and V_s can be established by using the Fermi-Dirac statistics:

$$Q_{ss} = eN_t \left\{ 1 - \frac{1}{1 + \exp\left[((E_t - E_F)_0 - eV_s)/kT\right]} \right\} \quad \text{for a donor state,} \tag{3.12}$$

$$Q_{ss} = -eN_t \frac{1}{1 + \exp\left[((E_t - E_F)_0 - eV_s)/kT\right]} \quad \text{for an acceptor state.} \tag{3.13}$$

N_t stands for the surface state density measured in states per unit area. $(E_t - E_F)_0$ is the energy difference between the gap state and the Fermi energy in the absence of the energy band bending.

With some further approximations the width w of SSCR in dependence of V_s can be deduced as [103]:

$$w = \sqrt{\frac{2\varepsilon_s V_s}{e|N_a - N_d|}} \tag{3.14}$$

ε_s is the dielectric constant and $|N_a - N_d|$ the net doping density. For CIGS absorbers, reasonable values $V_s = 0.4\,V$, $\varepsilon_s = 10\varepsilon_0$ [104] and $|N_a - N_d| = 10^{16}\,cm^{-3}$ [79] will result in $w = 210\,\text{nm}$. This value can give a rough estimation of the influence of the surface states on the bulk material – surface states localized within the very first several atomic layers can electrically affect a region of several thousands of atomic layers away from the surface.

Surface dipoles induced abrupt band offset

In addition to surface states, another important phenomenon associated with semiconductor surfaces is the surface dipole. As reviewed by Kronik *et al.* [103], at an ideal surface the wave functions of the electrons extend out of the surface. As a result, the region just outside the surface is negatively charged, whereas the region just inside the surface is positively charged as shown in Fig. 3.4 (b). The separation of positive and negative charges over several atomic monolayers forms microscopic dipoles creating an electrical field. This field repels electrons reaching the surface back into the bulk. In this way, an abrupt potential barrier $\Delta\phi$ for electrons attempting to leave the semiconductor is formed. Besides the spill out of the electron wave functions, the previously mentioned surface relaxation and reconstruction can also lead to microscopic dipoles. Affected by both the surface band bending and the surface dipoles, the work function at the surface can be described as $\phi_e = \phi_r \pm |e \cdot V_s| \pm |\Delta\phi_s|$. ϕ_r stands for the energy difference between the Fermi energy E_F in the bulk of the material and the vacuum energy E_{vac} far away from the surface. Depending on the type of the surface states (donor or acceptor) and the origin of the surface dipoles, the signs of V_s and $\Delta\phi_s$ can be positive or negative.

Surface photovoltage

The photovoltaic effect is in general an illumination-induced change in the potential distribution of a certain structure. A specific case of this effect is the surface photovoltaic effect that was firstly reported on Si in 1947 [105].

The surface photovoltage (SPV) is by definition the illumination-induced change in the surface potential distribution. Under illumination the absorbed photons generate free carriers by creating electron-hole pairs via band-band transitions (for super-band-gap photons) and/or releasing captured carriers via trap-band transitions (for sub-band-gap photons) [103]. Driven by the electrical field in the SSCR indicated in Fig. 3.5 (b), photo-generated electrons and holes are separated and induce an opposite electrical field to that in the SSCR. This field screens the surface band bending and changes the work function to a new value ϕ_i. The difference $\phi_e - \phi_i$ is the surface photovoltage SPV. Under adequate illumination the energy band bending can be theoretically totally compensated, which is called the photosaturation effect. This effect has found wide applications in investigating

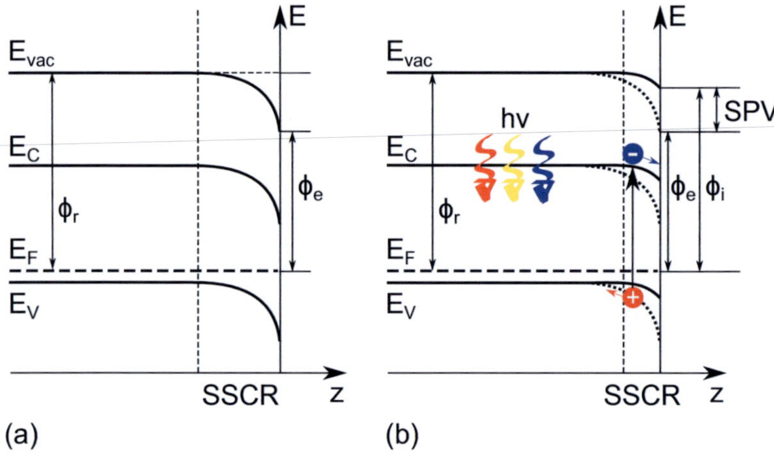

Figure 3.5: **(a)** The energy band diagram of a p-type material with the work function ϕ_e on the surface. **(b)** Under illumination free charge carriers are generated. Driven by the electrical field in the surface space charge region SSCR, the electrons and holes are separated and induce an opposite electrical field to that in the SSCR. Consequently, the surface band bending and the width of SSCR are reduced. Correspondingly, the work function on the surface is changed to ϕ_i. The difference $\phi_e - \phi_i$ is the surface photovoltage (SPV).

the surface band bending of various materials [106–110] and provides a good approximation of the properties of the bulk material.

3.3.2 Effect of adsorbates, oxidation and surface water layer

So far, the situation on ideally clean semiconductor surfaces in contact with perfect vacuum was discussed. However, such a condition can never be realized in the real world. In reality there are always impurity atoms and molecules adhering on the surfaces under study, which are called adsorbates. The electron wave functions of the adsorbates will extend into the bulk of the material and generate surface states, in complete analogy to the behavior of intrinsic surface states on ideally clean surfaces. Including the charge in the tails of their electron wave functions, adsorbates are electrically neutral. Depending on the character of these tails, the gravity center of the charges may shift towards or away from the surface, which will induce polarization or, in other words, surface dipoles [111]. Previously, the influence of adsorbate layer on the surface band bending of III-IV semiconductors [112–114], silicon [115–117], chalcogenide semiconductors [118–121] and particularly CIGS [122–125] has been extensively studied. Further research elucidated that not only inorganic atoms but also organic molecules can effectively change the surface work function of semiconductors. Making use of this feature, the electrical properties of surfaces can be even deliberately designed [126–132].

In addition, the oxygen-induced surface band bending was verified on various semiconductor surfaces [133–136]. Particularly on CIGS surfaces, Heske *et al.* reported the formation of the native oxide SeO_2 by air exposure [137]. Rau *et al.* showed the oxygenation-induced passivation of Se vacancies leading to an alteration of the surface band bending [138].

Last but not least, water films are generally present on solid surfaces in air. In the two review articles Refs. [139, 140] the fundamental aspects of this issue were explicitly discussed. Particularly for KPFM measurements in air, the water films can be easily polarized by applying the dc-voltage between the tip and sample. The polarization of the water films is opposite to that of the dc-voltage and will thus cancels the dc-voltage [141]. In an exemplary work established by Sugimura *et al.* [142], the phenomenon that water films shield the potential contrast was clearly shown. In that work, Si samples with p- and n-doped regions were prepared. By exposing a sample to vacuum ultraviolet light, the sample surface was overall terminated with hydroxyl (OH) groups. Under this condition, the potential contrast vanished. On the contrary, after the sample was annealed in air at a temperature of $100\,^{\circ}C$, a clear contrast appeared between differently doped areas. Similarly, a potential contrast could also be clearly resolved under a low humidity.

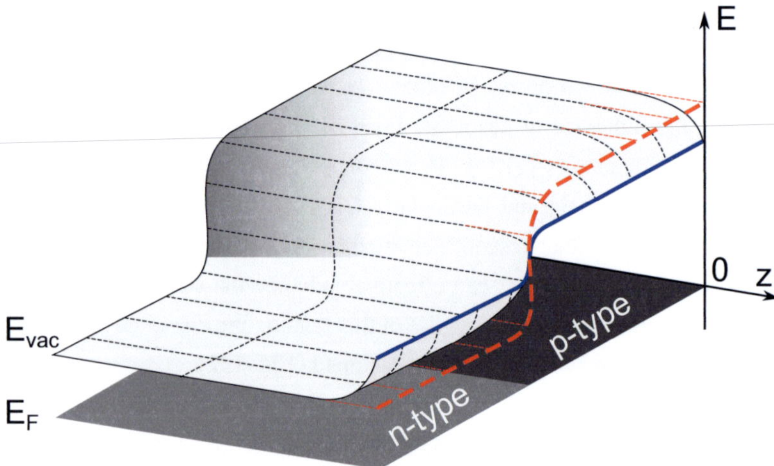

Figure 3.6: Influenced by surface effects like surface adsorbates, surface oxidation and water films between the tip and sample in ambient conditions, the desired value of the work function contrast of a pn-junction (red dashed line) is usually significantly reduced to an effective value on the surface (blue solid line). Therefore, the experimentally determined CPD contrast, which is proportional to the work function contrast, is usually noticeably smaller as expected.

By the joint action of all surface effects, the modification of the potential contrast on the surface of a semiconductor pn-junction is depicted in Fig. 3.6. In this illustration, surface band bending arises on both sides of the pn-junction. The desired value of the work function contrast (red dashed line) is usually significantly reduced to an effective value on the surface (blue solid line). Accordingly, the CPD contrast, which is proportional to the work function contrast, is usually noticeably smaller as expected from the bulk of the materials. Consequently, during the execution of the KPFM measurements and in the analysis of the CPD values the impact of the surface effects should be always carefully considered.

3.3.3 Effect of scattered laser beam

The main target of this work is to investigate the potential distributions on cross sections of CIGS solar cells in darkness and under controlled illumination. Therefore, we have to figure out at first, how "dark" is the darkness in measurement setup. As it will be introduced in more detail later on, a laser diode is reflected on the back side of the cantilever for monitoring the movement of the cantilever. Usually, the laser spot has a diameter of 40–50 μm and is larger than the width of the cantilever that is $10\,\mu$m. In the experimental setup of the current work, this can be verified by observing the scattered laser spot on the sample surface in comparison to the cantilever in Fig. 3.7 (a). However, in a dimension of several tens of micrometers, the laser beam can be observed as straight lines. Thus, the area right under the tip, or in other words, the area just being scanned is actually shadowed by the cantilever. Yet, the diffusion of the photogenerated charge carriers from the surrounding areas can still have an influence on this area. In order to verify this influence, a KPFM measurement was performed on the cross section of a CIGS solar cell without external illumination as shown in Fig. 3.7 (b). During the measurement the photovoltage between the ZnO and Mo electrodes was measured. It came out that the photovoltage V_{Photo}

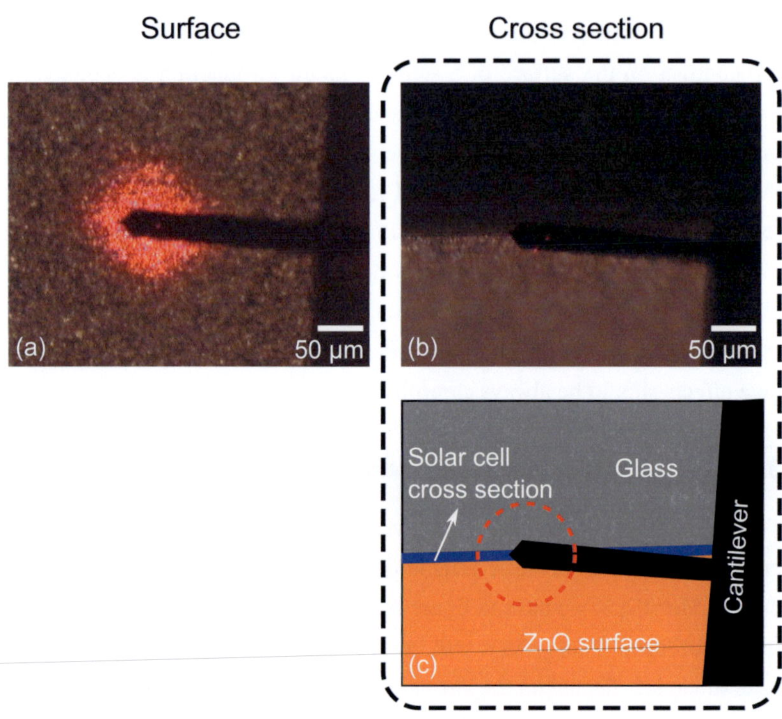

Figure 3.7: Photographs of the cantilever **(a)** on a CIGS absorber surface and **(b)** on a cross section of a CIGS solar cell. For clarity, the photograph (b) is explained with a sketch **(c)**. As clearly shown by the scattered laser light in (a), the radius of the laser spot is larger than the width of the cantilever. The approximate coverage range of the laser spot is indicated in (c) with a red dashed circle. Obviously, the cross section of the solar cell, shown with a thick blue line in (c), is exposed to the scattered laser light.

has a value of over $200\,\text{mV}$. As shown in Fig. 2.3 in the former section, the potential drop ΔCPD through the heterojunction will be reduced by V_{Photo} under illumination, i.e., $\Delta CPD_{illumination} = \Delta CPD_{dark} - V_{Photo}$. This means, the potential contrast between CIGS and ZnO is reduced by over $200\,\text{mV}$ due to the scattered laser light, additionally to all surface effects. Nevertheless, the sum $\Delta CPD + V_{Photo}$ can provide a good estimation of the real value of the potential drop induced by the solar cell heterojunction. This potential drop is the theoretical upper limit of the open circuit voltage V_{oc} and is therefore of great importance. An extensive discussion about this part can be found in Chap. 7.1.

3.4 Survey and discussion of previous studies

Important previous studies of KPFM on the surface of CIGS absorbers and on polished cross sections of CIGS solar cell devices are summarized in the following. The drawbacks in these studies are subsequently discussed, in order to introduce the improvement and breakthroughs in the current study.

3.4.1 Surfaces of Cu(In,Ga)(S,Se)$_2$ absorber layers

Many KPFM measurements on CIGS thin films were published by Sommerhalter *et al.* [47, 143]. In these studies, KPFM measurements were carried out in ultra-high vacuum (UHV) on CuGaSe$_2$ thin layers, which were epitaxially grown on GaAs(001) wafers. After upgrading this experimental setup, Sadewasser *et al.* analyzed the surface of two CuGaSe$_2$ thin layers grown epitaxially on a ZnSe(110) substrate and on a Mo-coated glass substrate. High potential contrast as large as 255 mV was found between different facets of single grains. This was probably the earliest work presenting clear potential distribution on the surface of a polycrystalline CIGS thin film. In a further study of Sadewasser *et al.* [144] on polycrystalline CuGaSe$_2$ and Cu(In,Ga)S$_2$ absorber materials, downward band bending of 110 meV was observed at individual grain boundaries, which was attributed to charged defect states at the interface between grains. A decrease of this band bending under illumination revealed a reduction of the potential barrier that limits the charge carrier transport across the grain boundaries. Later, on basis of similar results achieved on a CuIn$_{0.7}$Ga$_{0.3}$Se$_2$ sample in air Jiang *et al.* [145] claimed that the grain boundaries are positively charged and the local built-in potential at grain boundaries is expected to increase the collection of the minority carriers (electrons). In order to support this conclusion, the authors compared the local built-in potential in samples with [Ga]/([Ga]+[In])-ratios from 0 to 1 and the efficiencies of these samples. The results showed a strong correlation between these two quantities [13].

Shortly after that, Marrón and Sadewasser *et al.* [146–148] studied in UHV the rear side of a $CuGaSe_2$ sample peeled-off from the Mo substrate. Different SPV values were found at different grain boundaries demonstrating the existence of different types of grain boundaries. These grain boundaries were presumed to associate with different crystallite orientations. Interestingly, only one type of work function variation, i.e., the drop of work function, was reported at grain boundaries. This effect was considered to increase the current collection in the solar cell device, thereby compensating negative effects of recombination through defects at grain boundaries [149]. A variety of grain boundaries was observed by Hanna *et al.* [150], who showed a dip in the work function at grain boundaries in a randomly or (112)-textured $CuIn_{0.7}Ga_{0.3}Se_2$ sample and in contrast a step or spike in (220/204)-textured ones . Baier *et al.* was able to associate the symmetry of the grain boundaries to their electrical properties in a polycrystalline $CuInSe_2$ sample. In more detail, the potential barrier at grain boundaries was observed to be positive, negative, or zero. In these grain boundaries, the $\Sigma3$-type has a higher probability to be charge neutral than non-$\Sigma3$ ones. The observation of three types of potential variations on the same sample surface concurs with another work published by Sadewasser *et al.* on a $CuIn_{0.67}Ga_{0.33}Se_2$ sample [151]. Indeed, if one takes a closer look at the data in some former work, e.g., Fig. 1 of Ref. [144], it can be seen that there are more than one type of potential variation at grain boundaries. However, during that time the variety of grain boundaries did not gain further attention.

On basis of these observations the superior performance of polycrystalline $CuIn_{0.7}Ga_{0.3}Se_2$ solar cells could not be explained with the model based on single type of grain boundaries any more. To explain this, another model was brought forward, namely, the electrical neutrality of the grain boundaries. Since $\Sigma 3$ grain boundaries are the ones appearing frequently in the most successful solar cells, single $\Sigma 3$ grain boundaries were deliberately formed in an epitaxially grown $CuGaSe_2$ sample. The results combined from Hall measurement and KPFM indicated that $\Sigma 3$ grain boundaries are neutral barriers for holes with a barrier height of $30 - 40\,mV$ [73]. Yan *et al.* reported a first-principles calculation for $CuInSe_2$ material and confirmed the theoretical estimations on a $Cu(In,Ga)Se_2$ polycrystalline sample with KPFM results. The authors came to the conclusion that grain boundaries in $Cu(In,Ga)Se_2$ are electrically benign and the Na segregation on the grain boundaries is responsible for the potential variations [152] .

Not only the electrical properties of the absorber surface but also those of the absorber/buffer interface were proven to be crucial for the solar cell performance. For studying the impact of the CBD process on the interface formation, Jiang *et al.* [153] carried out KPFM measurements on the as-grown surface of a $Cu(In,Ga)Se_2$ absorber and after chemical treatment with solutions including the pure water, the water solution of NH_4OH and the water solution of $NH_4OH + CdSO_4$. With these measurements the electrical modification of the $Cu(In,Ga)Se_2$ surface before the formation of CdS was studied, leading to the conclusion that this modification can facilitate the junction formation of the device and is expected to be favorable for the device performance. Glatzel and Rusu *et al.* [154, 155] investigated the influence of CdS with thickness of a few nanometer on the work function distribution along grain boundaries of the absorber. The CdS films were grown on top of Se-decapped and air-exposed $CuIn_{0.76}Ga_{0.24}Se_2$ absorbers by physical vapor deposition (PVD). The results showed a pronounced decrease of the work function around the grain boundaries of Se-decapped absorbers suggesting a change of neutral grain boundaries to positively charged ones. This effect was ascribed to the diffusion process most likely of sulfur along the grain boundaries. Due to the oxide layer and adsorbates on the surface of the air-exposed absorber, the diffusion process is hampered.

Moreover, combining the surface photovoltage spectroscopy (SPS) with KPFM the optical properties with high spatial resolution can be analyzed. Streicher *et al.* [156, 157] employed this technique on a $CuGaSe_2$ thin film grown on GaAs substrate and $CuInS_2$, Zn-doped $CuInS_2$ thin films grown on Mo-covered glass substrate. The band gap energies within the grain structure were successfully extracted showing an interesting application of this technique.

3.4.2 Polished cross sections of Cu(In,Ga)(S,Se)$_2$ solar cell devices

Since the invention of KPFM in 1991 [87], this technique has been employed for studying potential distributions in semiconductor structures with different complexity, from single pn-junctions [158–160] with dimensions of micrometers to heterostructures [161–166] as small as several tens of nanometers. Applying this technique, even cross sections of operating devices like light-emitting diodes [167, 168], lasers [169, 170] and transistors [171–175] could be thoroughly investigated.

In particular for applications on solar cells, KPFM also unveiled to be an exceptional tool for two-dimensional potential imaging on cross sections of devices based on epitaxially grown III-V-semiconductors [176, 177] and multicrystalline materials, e.g., silicon [178], CdTe [179] and Cu(In,Ga)(S,Se)$_2$ [15, 180–183]. For measurements on devices based on epitaxially grown materials [159–174, 176, 177], KPFM could be simply performed on cleaved cross sections of the devices benefiting from the small roughnesses of a few nanometers. However, it is much tougher to study cleaved cross sections of devices consisting of multiple polycrystalline thin layers. The main reason is that large height differences between the broken edges of the layers are usually found after cleavage. These height differences can be larger than several ten micrometers and even exceed the measurement range of the piezo element in a KPFM setup. In order to overcome this issue, considerable efforts are necessary. This explains why there are only limited work successfully conducted on cross sections of Cu(In,Ga)(S,Se)$_2$ solar cells, in contrary to a great number of publications on absorber surfaces.

A pioneer work on this topic was conducted by Glatzel *et al.* [180, 184]. Their study published in 2002 was probably the first one about KPFM measurements on cross sections of all kinds of polycrystalline solar cells. In that work KPFM under UHV was used to image the electronic structure of a Mo/CuGaSe$_2$/CdS/ZnO thin-film solar cell as shown by Fig. 3.8. A sufficiently flat cross section was prepared by gluing two devices face-to-face and polishing with an aluminum paste. In order to clean the adsorbates and reduce the surface defects on the cross section, the sample was annealed at $100\,^\circ C$ for 1 hour followed by soft sputtering with argon

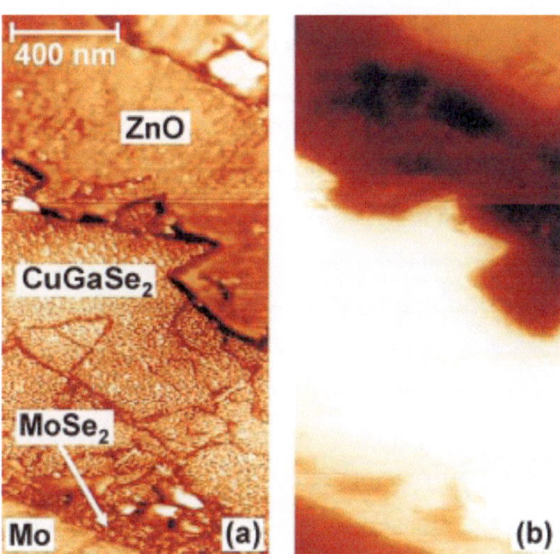

Figure 3.8: KPFM measurements on a polished cross section of a CuGaSe$_2$ solar cell with **(a)** topography and **(b)** CPD images. The cross section was cleaned by soft ion sputtering for 60 min. (Reproduced from Ref. [180])

(Ar) ions. The cleanliness of the cross section was checked by observing the work function contrast between p-type $CuGaSe_2$ and n-type ZnO, showing an optimum surface condition at a sputter time of 15 minutes. Moreover, a p-type $MoSe_2$ layer was resolved between the absorber and the Mo back contact. Following the same procedure Marrón *et al.* [181] observed different secondary phases in Cu-rich $CuGaSe_2$ absorbers, including the p-type degenerate $Cu_{2-x}Se$ phase. The SPV results showed a constant work function of $Cu_{2-x}Se$, which was interpreted as a fingerprint of the metallic character of this phase. In a further work of Glatzel *et al.* [182], the work function distributions in $Cu(In,Ga)(S,Se)_2$/i-ZnO/ZnO:Ga and $Cu(In,Ga)(S,Se)_2$/(Zn,Mg)O/ZnO:Ga heterostructures were studied. It seems that the position of the junction was moved by about 30 nm towards the absorber through substituting the i-ZnO layer with $Zn_{1-x}Mg_xO$. This outcome showed that the deposition of the (Zn,Mg)O layer can extend the inversion of the absorber surface. As a consequence, the pn-transition was moved further into the absorber layer leading to a better performance of the solar cell. Mainz *et al.* [183] applied KPFM on a cross section of a $Cu(In,Ga)S_2$/CdS/ZnO solar cell. The $Cu(In,Ga)S_2$ absorber layer in the solar cell was fabricated by a sequential process. Based on the potential distribution a clear $CuGaS_2$/$CuInS_2$ double layer structure with $CuGaS_2$ near the Mo/absorber interface was observed.

The only work performing KPFM on cleaved multicrystalline solar cells abandoning the polishing process was established by Jiang *et al.* [15], who fabricated $Cu(In,Ga)Se_2$ solar cells on a GaAs(001) wafer. One solar cell was cleaved along the $[1\bar{1}0]$ direction of the wafer, so that the cross section was flat enough for KPFM measurements. With the results presented in Fig. 3.9 the authors demonstrated that the pn-junction in $Cu(In,Ga)Se_2$ solar cell is a buried homojunction with the p/n boundary located 30 – 80 nm from the $Cu(In,Ga)Se_2$/CdS interface. Based on the observation that the electric field terminates at the $Cu(In,Ga)Se_2$/CdS, it was concluded that the CdS and ZnO layers are inactive for the collection of photoexcited carriers.

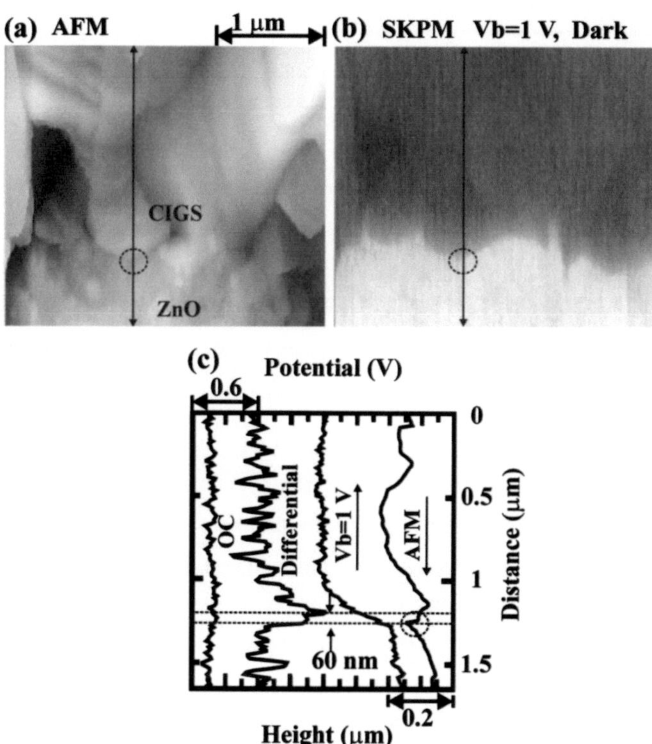

Figure 3.9: KPFM measurements on a cleaved cross section of a Cu(In,Ga)Se$_2$ solar cell grown on a GaAs(001) wafer. **(a)** and **(b)** display the topography and CPD images. **(c)** shows example topography and CPD line sections from this area. As shown by the line section OC, the CPD signal does not show any contrast in the open circuit condition of the solar cell. The CPD contrast shown in (b) was achieved with an external reverse bias of 1 V on the solar cell. (Reproduced from Ref. [15])

Unfortunately, the Fermi level pinning seems to have a considerable influence on the surface of the cross section in that work. As a consequence, the measurement in darkness under open circuit condition exhibited nearly no potential contrast and the aforementioned statements were made with a reverse bias of 1 V. Furthermore, the surface photovoltage did not saturate until an illumination intensity of $230\,\mathrm{mW/cm^2}$ and the potential drop at the pn-junction disappeared again at an intensity of $360\,\mathrm{mW/cm^2}$, which is a multiple of the intensity of the standard test conditions ($100\,\mathrm{mW/cm^2}$). Nevertheless, this exploratory work showed the possibility to investigate cross sections of this kind of solar cell without modifying its topographic structure.

3.4.3 Discussion about the drawbacks in previous studies

Surfaces of $Cu(In,Ga)(S,Se)_2$ absorbers

Until now, most electrical models for CIGS grain boundaries were developed based on KPFM results on surfaces of CIGS absorbers. However, it has been published for two decays that the surface of CIGS absorbers often has a different material composition (In-rich in Ref. [185], Cu-poor in Ref. [186, 187]) and even a type inversion (n-type) [185] in comparison with the bulk material. This surface layer is usually referred to as the ordered vacancy compound (OVC) and may have a chemical compound of $CuIn_3Se_5$ [185, 188] or even other forms such as $CuIn_5Se_8$ and $Cu_2In_4Se_7$ as suggested by simulation results [189]. As it is well known, KPFM is a highly surface-sensitive measurement technique [190]. Consequently, it is actually inappropriate to interpret the functionality of the grain structure on basis of results from the surfaces. The ultimate way to resolve the puzzle is certainly to look at the grain boundaries buried in the bulk material directly. Due to the 3-dimensional distribution, grain boundaries in the bulk should form the majority compared to those on the absorber surface (2-dimensional), and are thereby more decisive for the solar cell performance. Unfortunately, there are hitherto no such measurement techniques available for this task without exposing the buried grain boundaries. Thus, practically the best approximation is to investigate grain boundaries on cleaved cross sections of the absorbers. Despite the surface effects, properties of grain boundaries on untreated cross sections will be very close to those of the buried ones (at least more closely as the ones on the absorber surface). This consideration led the author of this work to compare grain boundaries on the surface and on cleaved cross sections of the same CIGS absorber in Chap. 6. The results verify remarkable differences between grain boundaries from these two positions and provide a more general interpretation of the functionality of the CIGS grain structure.

Polished cross sections of Cu(In,Ga)(S,Se)$_2$ solar cell devices

As previously mentioned, by employing a mechanical polishing process cross sections sufficiently flat for KPFM measurements can be prepared. However, two problems arise with the polishing process.

First, the original properties of the cross sections are inevitably changed. Again, KPFM is highly surface-sensitive, the modification of the surfaces will have significant influence on the interpretation of the measurement data. In order to minimize the modification, in some works the polished cross sections were cleaned by soft sputtering with Ar ions. In this way an increase of the potential contrast between the layers could be observed [180, 182]. Unfortunately, the situation cannot be totally retrieved. For example, it is unclear, whether the texture in the topography image of Fig. 3.8 stemmed from the grain structure or simply structural damages caused by the polishing process. This makes an analysis of the potential distributions of the chalcopyrite grain structure in most of these works [180, 182, 183] a "mission impossible".

Second, to prevent the layer stack from separating from the substrate during the polishing process, twin samples were usually in use and glued face-to-face on their top electrodes. In this way, it was impossible to illuminate the sample from their top transparent electrode as one normally does for operating solar cells. Instead, an external bias was varied to manipulate the potential distribution at cross sections of polycrystalline solar cells [178, 179]. However, illuminating the solar cells has undoubtedly a great significance for understanding their real properties, since the performances of these devices are exclusively determined under illumination.

Given all the above points, it is clear that the best way to achieve possibly original properties of the solar cell is to performance KPFM measurements on untreated, or in other words, as-cleaved cross sections. The following chapter shows, how the technical breakthrough for preparing untreated cross sections of CIGS thin-film solar cells was realized.

4 Experimental details

In this chapter the experimental setup of the KPFM measurement is explicitly introduced. The relevant parameters in the setup are optimized on a reference sample. Moreover, the preparation procedure for untreated cross sections of CIGS solar cells is described, which enables KPFM measurements in the upcoming chapters.

4.1 Experimental setup

Schematic diagram

Figure 4.1 shows the schematic diagram of the experimental setup. It is depicted based on Ref. [191] published by Ziegler *et al.* The whole setup can be briefly divided into the AFM head, the scanner and four controllers: Oscillation Control 1 (OC1), Oscillation Control 2 (OC2), Z-Controller (ZC) and Kelvin Controller (KC). As introduced in the last chapter about the principle of the single mode KPFM, the tip-sample system is modulated with two ac-voltages $V_{ac}(f_0)$ and $V_{ac}(f_1)$. $V_{ac}(f_0)$ is provided by the Oscillator 1 and is applied on the shifter mounted at the end of the cantilever holder, in order to generate a mechanical vibration of the cantilever. $V_{ac}(f_1)$ is provided by the Oscillator 2 and is applied between the tip and sample, in order to modulate the electrostatic force. The movement of the cantilever is monitored by reflecting a laser beam on the back side of the cantilever and detecting the reflected beam with a four-quadrant-photodiode. The output of the photodiode is digitalized and fed to OC1 and OC2.

With a low-pass (LP) filter and a lock-in amplifier with the reference frequency f_0 provided by Oscillator 1, OC1 filters out the oscillation ampli-

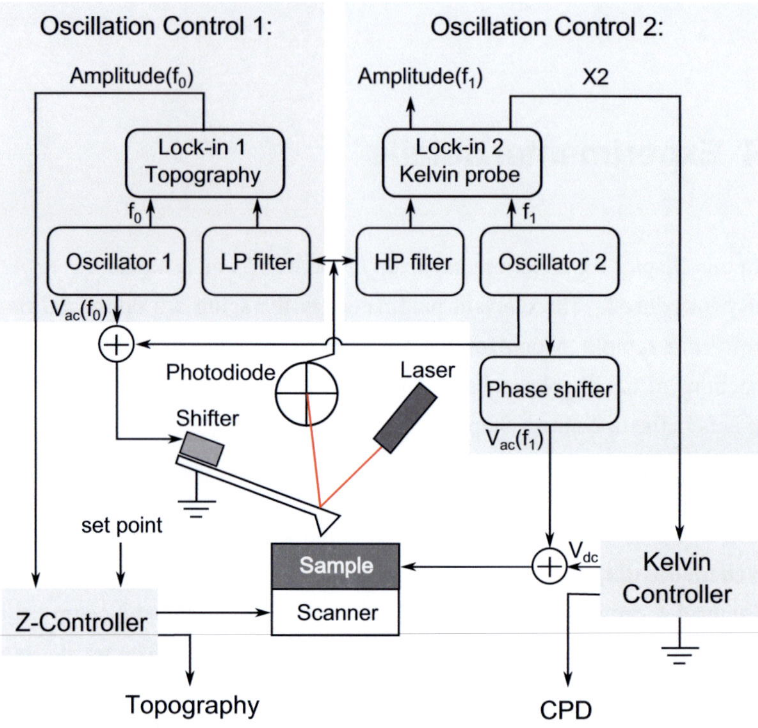

Figure 4.1: A schematic illustration of the experimental setup for single mode Kelvin probe force microscopy employed in this work. The setup consists mainly of 4 controllers: Oscillation Control 1, Oscillation Control 2, Z-Controller and Kelvin Controller. These four controllers are indicated with four gray blocks.

tude at f_0. This amplitude signal is given to ZC and compared to the set point. Based on the comparison result, ZC adjusts the Z-axis of the scanner, until the set point is reached. At this moment, the value of the Z-axis is recorded as the topography signal. Similarly, with a high pass (HP) filter and lock-in amplifier referred with f_1 from Oscillator 2, OC2 filters out the signal at f_1. As the input for KC, both the oscillation amplitude at f_1 and the X2 signal, which is a projection signal of the oscillation amplitude (more details are described in Appendix B), can be used. For convenience X2 is supplied to KC, which applies a dc-voltage V_{dc} to the sample surface. KC varies V_{dc}, until the magnitude of the X2 signal is nullified. That value of V_{dc} is documented as the CPD signal.

Multi-Mode AFM head

Figure 4.2 is a structural drawing of the measuring head of the Veeco MultiMode^{TM} AFM. As shown in the drawing, the laser beam (1) is successively reflected at a wedge mirror (2), the back surface of the cantilever (3) and a tilt mirror (4) for three times and finally detected by the four-quadrant-photodetector (5). The position of the laser beam relative to the cantilever can be adjusted by knobs (6). The facing angle of the photodetector is adjustable with knobs (7). The position of the cantilever together with the laser beam can be changed relative to the sample surface with knobs (8). Moreover, on the back side of the AFM head (not shown in the

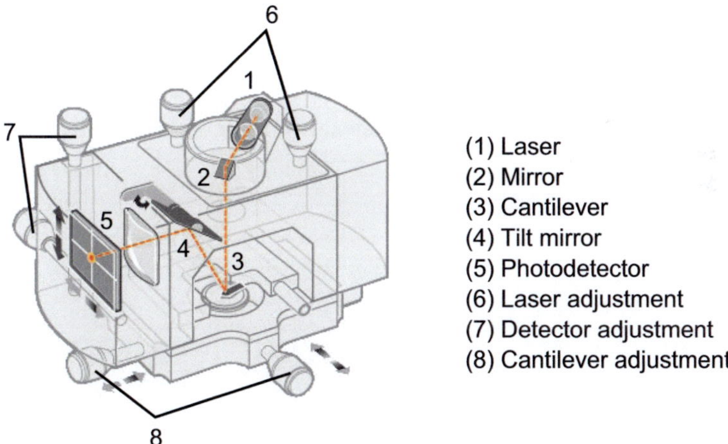

Figure 4.2: A structural drawing of the AFM head on basis of the drawing in Ref. [192].

drawing) there is a clamp that fixes and conducts the cantilever holder, and a lever responsible for tilting the mirror (4). Every time if a new cantilever is exchanged, the laser beam has to be adjusted with knobs (6) until the laser spot is in length direction at the front part and in width direction in the middle of the cantilever. As the next step, the photodetector will be adjusted with knobs (7), until the reflected laser beam hits the center of the photodetector. If a special position on the sample surface is wished to be scanned, the cantilever can be driven to the destination with knobs (8). Normally, mirror (4) stays unchanged during the calibration.

Cantilever

The cantilevers used in the present work are of the type PPP-FMR from NANOSENSORS^{TM} [193]. These cantilevers are made of highly doped silicon. Some SEM images of one cantilever are shown in Fig. 4.3. The specifications of this type of cantilever are listed in Tab. 4.1. In addition to the technical data according to the manufacturer, the measured height of the apex is $10\,\mu$m and the second resonance frequency is around $450\,$kHz. Furthermore, the rear side of the cantilever is coated with $30\,$nm aluminum. This reflex coating enhances the reflectivity of the laser beam by a factor of about 2.5. Therewith, the interference of the laser beam within the cantilever, which deteriorates the detection quality, can be effectively reduced.

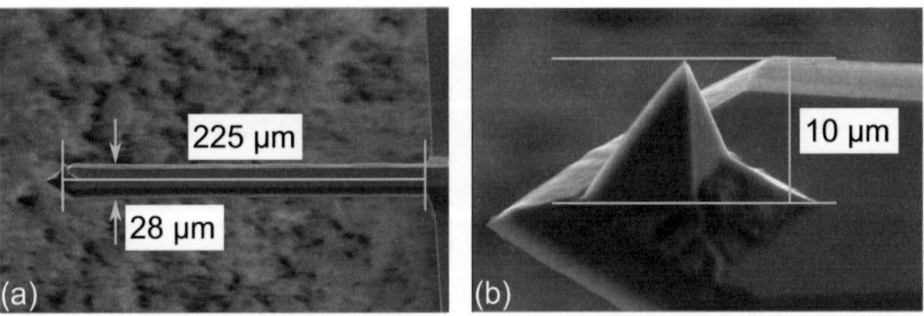

Figure 4.3: SEM images of a cantilever of the type PPP-FMR.

Length	Width	Thickness	Force constant
$225\,\mu$m	$28\,\mu$m	$3\,\mu$m	2N/m

Resonance frequency	Tip radius
$75\,$kHz	$<7\,$nm

Table 4.1: Specifications of the cantilevers of the type PPP-FMR from NANOSENSORS^{TM}.

Scanner

The scanner in use is an iC AFM Upgrade Kit (nPoint, Inc.) for the Veeco MultiMode^{TM} AFM. It has a scan range of $100 \times 100 \times 15\,\mu$m (X, Y and Z). The noticeable increase of extension range in Z direction ($6\,\mu$m with the original scanner) is a key factor enabling the measurement on very rough cross sections. The scanner is controlled by a C.300 DSP controller based on a digital signal processor (DSP). This control system can provide closed-loop operation for all three scan directions X, Y and Z. The operation parameters of the scanner and the control system can be found in Refs. [194, 195].

SPM Control System

The SPM Control System from Nanonis^{TM} is applied in the KPFM setup. A circuit diagram of the system and the functionalities of individual control units can be found in Appendix A. The whole system comprises an Adaptation Kit AKVM, a Dual-OC4 station out of OC4-1 and OC4-2, and a basic package consisting of a Signal Conditioning unit (SC4) and a Real-time Controller (RTC). The voltages are applied with BNC cables and the communication between the units is realized through serial ports.

The user interface of the software is designed corresponding to the schematic diagram in Fig. 4.1. Besides the aforementioned four modules Oscillation Control (OC), Oscillation Control 2 (OC2), Z-Controller (ZC) and Kelvin

Controller (KC), a fifth module the Scan Control (SC) is inserted. The calibration procedure by setting the related parameters in these modules can be found in Appendix B.

Halogen lamp

A halogen lamp (Dolan-Jenner, Fiber-Lite-PL800) is used to illuminate the samples with controlled intensities. The light beam is collimated by a collimator lens for higher illumination intensities. A high resolution spectrometer Spectro 320D of Instrument Systems GmbH is used to characterize the illumination intensities. The spectral irradiance at the defined levels are shown in Fig. 4.4. As a reference, the solar spectral irradiance of air mass (AM) 1.5 is also depicted. By integrating the spectral irradiance from

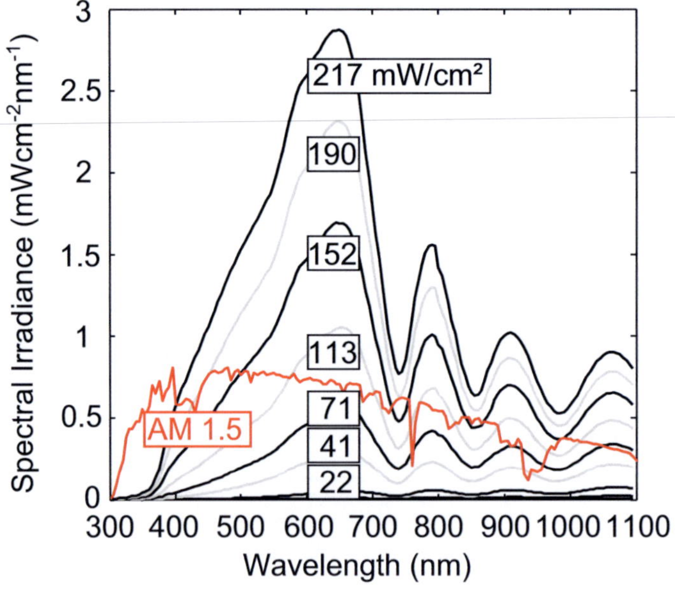

Figure 4.4: Spectral irradiance at defined levels of the halogen lamp. The solar spectral irradiance at air mass (AM) 1.5 with an intensity of $100\,\mathrm{mW/cm^2}$ is shown as the reference.

Level	I (mW/cm^2)	$I_{eff} = I \times \sin 45°$ (mW/cm^2)
1	1	0.7
2	5	3.5
3	22	16
3.5	41	29
4	71	50
4.5	113	80
5	152	107
5.5	190	134
6	217	153

Table 4.2: By integrating the spectral irradiance from 300 nm to 1100 nm, the illumination intensities are calculated. The tilt angle of the samples is taken into account by multiplying these values by a factor of $\sin 45°$.

300 nm to 1100 nm, the illumination intensities are calculated as listed in Tab. 4.2. Note that the samples are fixed on the magnetic pads with a tilt angle of about 45°. This tilt angle results in the effective illumination intensities for the samples, which are calculated as the measured intensities multiplying a factor of $\sin 45°$. In the following chapters the given illumination intensities are always the effective ones.

4.2 Sample preparation

Au/Mo reference sample

For testing the KPFM setup an Au/Mo reference sample is fabricated. A Mo-coated glass substrate for preparing CIGS solar cells is firstly cleaned in acetone and isopropanol. Subsequently, a grid (G2785C PLANCO GmbH) used for transmission electron microscopy is fixed on the Mo surface by gluing its edge with a silver paste. After that, a gold layer with the thickness of 100 nm is deposited by physical vapor deposition on the sample. Finally, the grid is carefully peeled off from the Mo surface. In this way, gold islands are formed on the surface of the Mo-coated glass substrate.

Surface of Cu(In,Ga)Se$_2$ absorber layer

Surfaces of CIGS absorber layers are achieved by etching complete solar cell devices with hydrogen chloride (HCl, 37%) for 60 seconds. In order to remove the remaining HCl on the sample surface, the samples are immersed in deionised water for 2 minutes and subsequently dried by a nitrogen pistol. After the etching process the ZnO layer and CdS layer are totally resolved and the CIGS surface is exposed. As reported by Liao *et al.* [196], CIGS is chemically resistent to HCl. In order to verify that the electrical properties of the CIGS surface is not influenced by HCl etching neither, both a fresh etched and an as-grown surface of a CIGS absorber are investigated with KPFM. The results from both surfaces do not show noticeable difference.

Untreated cross section of Cu(In,Ga)Se$_2$ solar cell

The process for preparing untreated cross sections is illustrated in Fig. 4.5. It can be described in the following steps: (a) a complete CIGS solar cell is first placed upside down on the preparation stage; (b) the rear side of the glass substrate is slit by a diamond cutter; (c) a glass plier (BO700, Silberschnitt $^{®}$) with gum pads on both sides of its beak is used to cleave the sample. In the middle of one gum pad there is a small bulge; (d) the slit of the glass substrate is put on the bulge. By gently pressing the plier, the solar cell is easily cleaved without generating extremely rough cross sections; (e) due to the limited space in the measurement head of the KPFM setup, the size of the cleaved solar cell has to be reduced. Therefore, one cleaved solar cell is glued on a target and cut by a wire saw (Model 3242, Well Diamond Wire Saws, Inc.). The cut is located about 2 mm away from

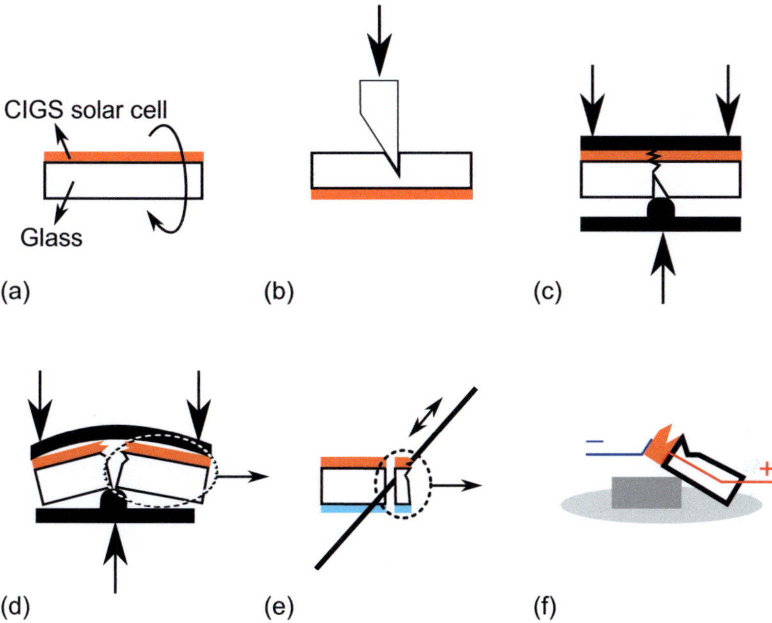

Figure 4.5: The preparation process for untreated cross sections of CIGS solar cells includes the following steps: **(a)** a complete CIGS solar cell is placed upside down on the preparation stage; **(b)** the rear side of the glass substrate is slit by a diamond cutter; **(c)** a glass plier with special design is used to cleave the solar cell; **(d)** by gently pressing the plier, the solar cell is easily cleaved; **(e)** the cleaved solar cell is then glued on a target. A small part including the cleaved cross section is cut off using a wire saw. **(f)** finally, this small part is fixed on a magnetic pad with a tilt angle around 45° and the electrodes of the solar cell are separately contacted.

the cleaved cross section. In this way the sample is cut into a thin piece that can be easily put into the measurement head. It should be pointed out that the rotational speed of the wire saw has to be adjusted slowly enough, so that the cooling water splashed by the wire saw does not contaminate the untreated cross section; (f) finally, the sample is fixed on a magnetic pad with a tilt angle around 45°, and the ZnO and Mo layers are individually conducted with copper tapes and silver paste. In all these preparation steps, the touch of the cross section should be carefully avoided.

Figure 4.6 is illustrated to clarify the different sample alignments for measurements on the CIGS absorber surface and on an untreated cross section of a complete solar cell device. It should be pointed out that the tilted sample alignment plays a central role in KPFM measurements on untreated cross sections. Compared to a perpendicular sample alignment ($\theta = 90°$) that one usually would adopt, a tilted sample alignment has two main advantages: first, the original height difference l between the layer edges becomes $l \cdot \sin\theta$. In this way the large high differences between all the layers in the solar cell and as a consequence the large roughness on untreated cross sections can be effectively reduced; second, with a perpendicular sample alignment, if the cantilever scans beyond the ZnO edge, it would fall abruptly into the air. Consequently, the measurement would have to break down. In the worst case the cantilever could be damaged. In contrast, with a tilted sample alignment, if the cantilever moves out from the ZnO edge, it will still stay on the top surface of the ZnO layer, and is thereby protected from damaging. Usually, with a scan area of $6 \times 6\,\mu m^2$ all the

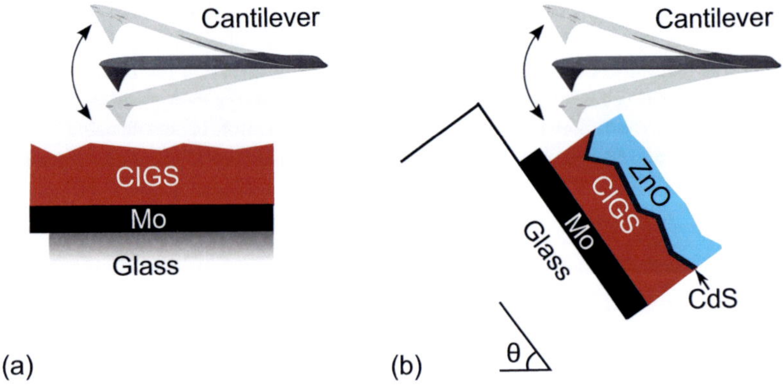

Figure 4.6: Sample alignments for KPFM measurements **(a)** on the surface of a CIGS absorber and **(b)** on the untreated cross section of a CIGS solar cell.

layers in the solar cell can be included and the height difference still does not exceed the measurement range in Z-direction of the setup.

4.3 Parameter optimization on the reference sample

As shown in the former section, there are plenty of adjustable parameters in the control panels of the two oscillation controllers and the Kelvin Controller. The most important two of them are the tip-sample distance and the amplitude of the ac-voltage $V_{ac}(f_1)$ that decide the resolution of the topography and the CPD images, respectively.

Tip-sample distance

As it was discussed in the last chapter, the cantilever oscillates without the interaction with the sample surface with a free oscillation amplitude at the first resonance frequency f_0. This free oscillation amplitude is proportional to $V_{ac}(f_0)$ applied on the shaker. The cantilever is then engaged to the sample surface, until the oscillation amplitude at f_0 is reduced to the set point, which correlates to a certain tip-sample distance. During the measurement the oscillation amplitude is kept constant at the set point for the topography imaging. As mentioned in the introduction of the contact, non-contact and tapping mode in Chap. 3.2.1, the tip-sample distance has obviously a great influence on the resolution of the topography image. In the case of KPFM, not only the resolution of the topography but also that of the CPD is of interest. Therefore, the influence of the tip-sample distance on both signals is carefully examined on the Au/Mo reference sample. Fig. 4.7 shows the topography and CPD line scans on a Au/Mo/Au structure with gradually reduced set point. The free oscillation amplitude is chosen as 350 mV and $V_{ac}(f_1)$ is set to 2 V.

Maybe it is noticed that the oscillation amplitude and the set point are given in "mV", instead of a length unit. The reason is that the oscillation amplitude is practically detected by the four-quadrant-photodetector. The output

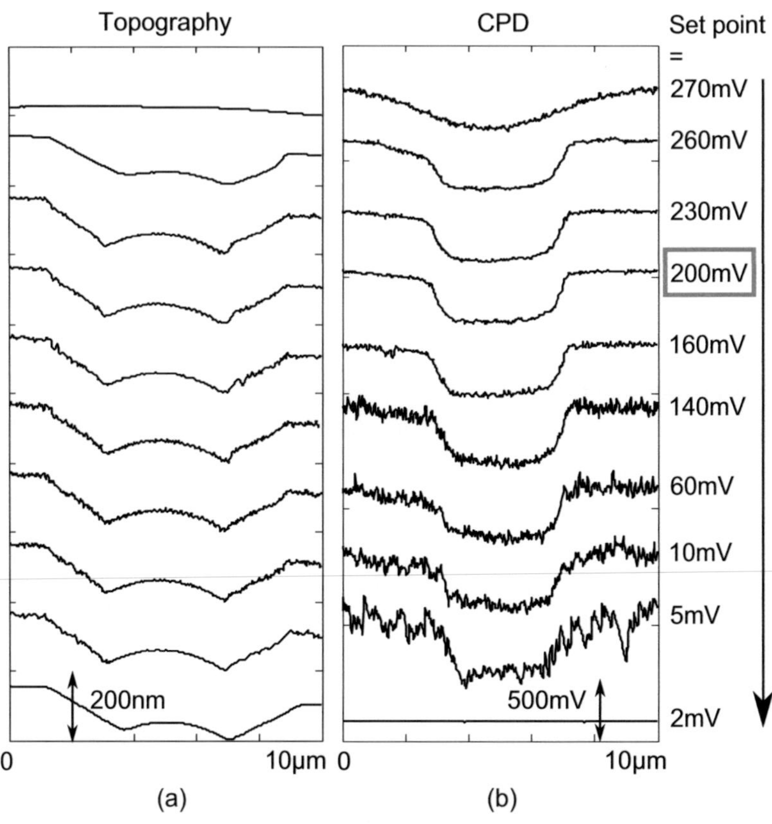

Figure 4.7: **(a)** The topography and **(b)** CPD line scans on an Au/Mo/Au structure under variation of the set point. The contrast in the CPD signal appears at 270 mV. By reducing the set point the contrast in both signals gradually improves. An optimum regarding the signal quality is observed at 200 mV. By further reducing the set point the topography signal does not change noticeably. However, the CPD signal becomes obviously much more noisy. At the set point of 2 mV the contrast of the CPD signal vanishes indicating that the tip already stabs on the sample surface at such a small distance. For these measurements $V_{ac}(f_1)$ is set to 2 V.

signal of the photodetector in "mV" shows the spatial deviation of the laser spot from the origin and is thus proportional to the oscillation amplitude of the cantilever. With the Z-Spectroscopy shown in Fig. 4.8 the signals in "mV" can be converted into lengths in "nm". For convenience, the original values in "mV" are used in the following discussion about the optimization of the set point.

As it can be observed in Fig. 4.7, at a set point of 270 mV the CPD contrast starts to appear. However, no variation in the topography signal is detectable. This is due to the long range nature of the electrostatic force in comparison to the van der Waals force [100, 197, 198]. By reducing the set point the contrast for both signals gradually improves. An optimum regarding the signal quality is found at the set point of 200 mV. According to the Z-Spectroscopy made afterwards, this set point corresponds to a tip-sample distance of about 13 nm (see Fig. 4.8). A further reduction of the set point has only a minor influence on the topography (the signal becomes even slightly less noisy), whereas the CPD signal turns to be obviously much more noisy down to a set point of 5 mV. At the set point of 2 mV the contrast of CPD totally vanishes with only two small pits at positions with structural changes. This observation can be explained as follows: by reducing the set point the tip-sample distance decreases, until the tip stabs on the sample surface. At such a small distance the cantilever is in principle operated in the contact mode and the oscillation at f_1 nearly vanishes. Consequently, the CPD is not detectable any more.

As just mentioned, with the Z-Spectroscopy shown in Fig. 4.8 the set point can be accurately determined. In the Z-Spectroscopy the cantilever approaches the sample surface. The free oscillation amplitude already decreases at distances of some micrometers. This is due to the additional damping of the cantilever by the air squeezed between the cantilever and the sample surface [199]. Consequently, the free oscillation amplitude at the position $Z_{relative} = 0$ is smaller than the aforementioned 350 mV. By further approaching the sample by 60 nm ($Z_{relative}$ from 0 to -60 nm) the

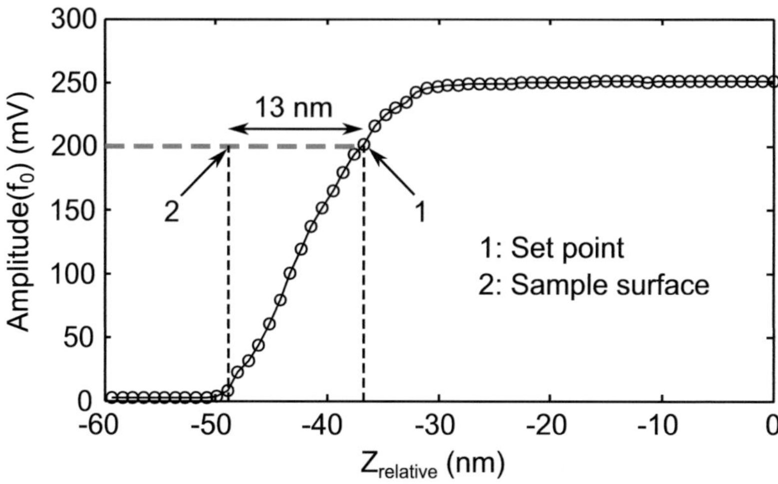

Figure 4.8: By approaching the cantilever to the sample surface and observing the variation of the oscillation amplitude at f_0, the Z-Spectroscopy is performed. At the position $Z_{relative} = -36$ nm the cantilever passes through the set point of 200 mV. At $Z_{relative} = -49$ nm the oscillation amplitude vanishes indicating a direct contact of the tip and sample. Consequently, the distance of 13 nm between these two positions is the averaged tip-sample distance for a set point of 200 mV. Depending on the cantilever and sample, the tip-sample distance for the same set point can differ slightly. Nevertheless, it stays stably within the range $10 - 15$ nm.

oscillation amplitude at f_0 is monitored. At the position $Z_{relative} = -36$ nm the cantilever passes through the set point and at $Z_{relative} = -49$ nm the oscillation amplitude vanishes implying a direct contact of the cantilever with the sample surface. Consequently, the distance of 13 nm is the averaged tip-sample distance at a set point of 200 mV. Note that depending on the cantilever and the sample the tip-sample distance for the same set point can slightly vary. However, it stays stably within the range $10 - 15$ nm.

Amplitude of $V_{ac}(f_1)$

At the optimal set point of 200 mV, the amplitude of $V_{ac}(f_1)$ is varied during a measurement on the same Au/Mo/Au structure. The results in Fig. 4.9

show clearly that $V_{ac}(f_1)$ has no influence on the topography signal. However, the signal quality of CPD is strongly dependent on $V_{ac}(f_1)$. At a value of 0.1 V the CPD signal is quite noisy. By increasing $V_{ac}(f_1)$ the signal quality gradually improves. The optimal signal quality is found in the range $1 - 2$ V. At even larger $V_{ac}(f_1)$ values the CPD difference between Au and Mo stays unchanged. However, the absolute values show an offset to the previous ones. This observation is most likely due to the surface band bending induced by large $V_{ac}(f_1)$, which was reported by Sommerhalter and Glatzel *et al.* [100, 200]. In their studies small values of $V_{ac}(f_1)$ were recommended for high measurement sensitivity. Consequently, the amplitude of $V_{ac}(f_1)$ between 1 V and 2 V is employed for the following KPFM study.

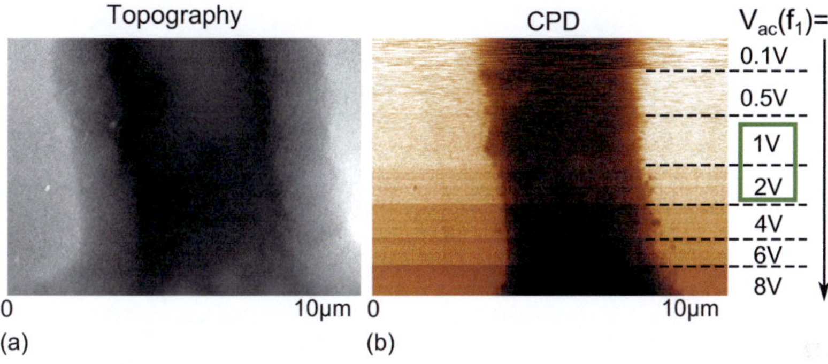

Figure 4.9: With the optimal set point of 200 mV, the variation of the (a) topography and (b) CPD line scans on the Au/Mo/Au structure in dependence of the ac-voltage at the second resonance frequency $V_{ac}(f_1)$ is observed. By increasing $V_{ac}(f_1)$ from 0.1 V to 2 V the signal quality gradually improves. At even larger values the CPD difference between Au and Mo does not change. However, the absolute CPD values show an offset to the previous one. This is most likely due to the surface band bending induced by large $V_{ac}(f_1)$. Consequently, $V_{ac}(f_1)$ between 1 V and 2 V is employed in this work.

4.4 Measurements on the reference sample

With the optimized parameter setting, i.e., the set point of 200 mV and $V_{ac}(f_1)$ of 2 V the Au/Mo reference sample is scanned by a size of $40 \times 40\,\mu m^2$. Fig. 4.10 (a) and (b) show the topography and CPD images. As it can be distinctly observed in the topography image, the Au islands have a form like the basis of a pyramid with a ramp on each side. However, in

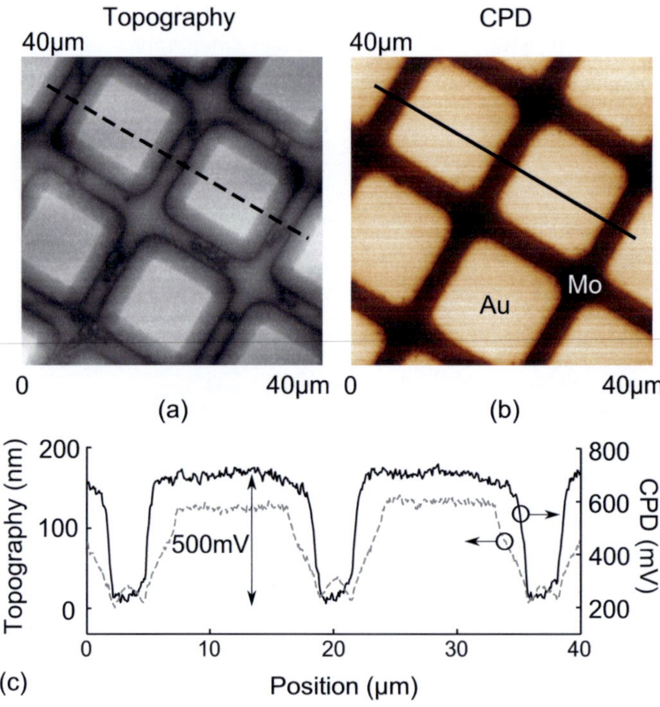

Figure 4.10: **(a)** Topography and **(b)** CPD images on the Au/Mo reference sample with a spatial resolution of 80 nm/pixel. The line sections in **(c)** are extracted at the position of the dashed and solid lines in (a) and (b). The results show that independent of the topography signal, CPD contrast can only be observed between different materials. This shows clearly that there is no convolution between the topography and CPD signals.

the CPD image only contrast between different materials can be observed, totally independent of the surface structure. This can be seen more clearly in the line scans shown in Fig. 4.10 (c). This important finding implies that there is no convolution between the topography and CPD signals, which is generally of the utmost significance for KPFM measurements [201, 202]. A closer analysis of the data shows that the height of the Au islands around 100 nm agrees well to the desired value. Moreover, the CPD difference between Au and Mo of 500 mV concurs nicely to literature values for the work function of these two metals (Au 5.1 eV and Mo 4.6 eV [203]). By now, the KPFM setup is successfully optimized and ready for the operation on CIGS solar cells.

5 Influence of the Ga content on the performance of $CuIn_{1-x}Ga_xSe_2$ solar cells

In this chapter four CIGS solar cells with varying Ga content are investigated with macroscopic (jV and EQE) and microscopic (KPFM) measurement methods. Based on these results, the Fermi energy shifting in the CIGS absorbers and different magnitude of charge carrier recombination in solar cell devices are studied.

5.1 jV-characteristics of $CuIn_{1-x}Ga_xSe_2$ solar cells

The jV-curves of solar cell devices fabricated on CIGS absorbers with x=[Ga]/([Ga]+[In])-ratios (GGI) of 0 ($CuInSe_2$), 0.32, 0.63 and 1 ($CuGaSe_2$) are shown in Fig.5.1. The significant photovoltaic parameters are extracted and listed in Tab.5.1. With an increasing GGI-ratio, the short circuit current density j_{sc} decreases while the open circuit voltage V_{oc} increases. An optimum for the fill factor FF and the power conversion efficiency η is observed at a GGI value of 0.32. Similar results were published over two decades ago [34] and have been verified by diverse research groups [9, 36, 37]. Note that the Ga content in CIGS absorbers studied in this chapter is deliberately kept constant during the growth, the benefit of the GGI gradient [204] has to be abandoned. As a result, the best power conversion efficiency in this sample series is below the ones that are usually achieved (η>16%). The chemical composition of the absorbers including the Ga cotent and the Cu content is determined by X-ray fluorescence (XRF) analysis. Both the jV-measurements and the XRF analysis are performed at ZSW.

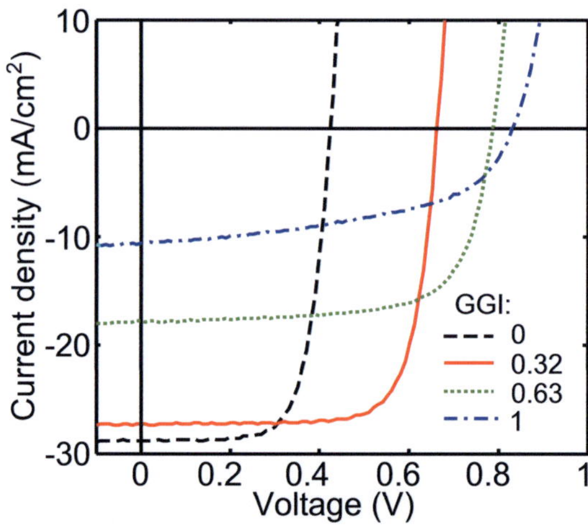

Figure 5.1: jV-curves of CIGS solar cells with different [Ga]/([Ga]+[In])-ratios (GGI). The significant photovoltaic parameters are listed in Tab.5.1 in the following. The jV-measurements are performed at ZSW.

GGI	Cu (at.%)	j_{sc} (mA/cm^2)	V_{oc} (mV)	FF (%)	η (%)
0	21.3	28.8	425	70.0	8.6
0.32	21.7	27.3	661	75.8	13.7
0.63	20.9	17.9	787	70.3	9.9
1	23.1	10.6	832	51.7	4.6

Table 5.1: Significant photovoltaic parameters from jV-measurements of CIGS solar cells with different GGI-ratios. The Ga and Cu content are determined by X-ray fluorescence (XRF) analysis at ZSW.

5.2 EQE-measurements of $CuIn_{1-x}Ga_xSe_2$ solar cells

In Fig. 5.2 (a) the external quantum efficiencies (EQE) of CIGS solar cells with different Ga content are exhibited. With an increasing Ga content the absorption edge of the solar cell device moves to shorter wavelengths cor-

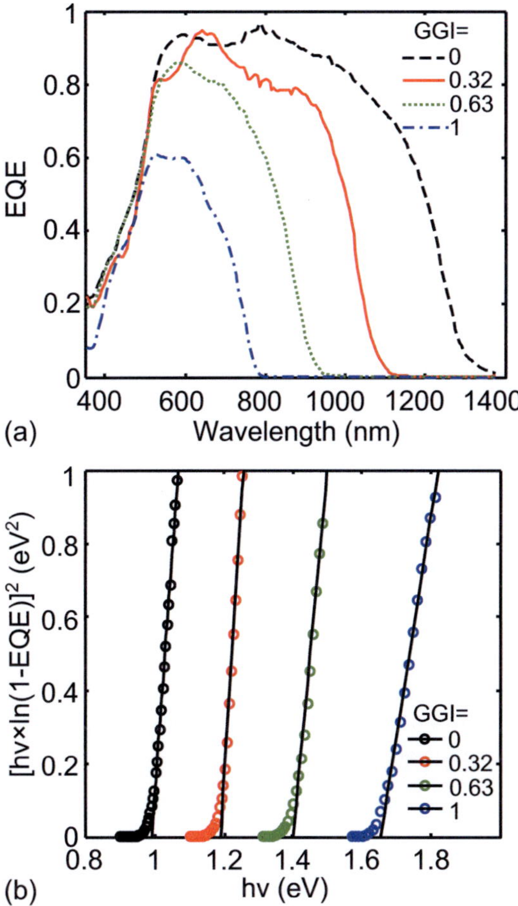

Figure 5.2: **(a)** External quantum efficiencies of CIGS solar cells with varying GGI-ratios. **(b)** By linearly fitting the function $\left[h\nu \times \ln\left(1 - EQE\left(h\nu\right)\right)\right]^2$ against the photon energy $h\nu$, the band gap energies of absorbers with increasing GGI-ratios are determined as 0.99, 1.19, 1.40 and 1.66 eV.

responding to higher optical energies. As shown in Fig. 5.2 (b), the measurement data near the individual band gap energies E_g are used to form $\left[hv \times \ln\left(1 - EQE\left(hv\right)\right)\right]^2$ as a function of hv. The band gap energies E_g determined for increasing Ga content are 0.99, 1.19, 1.40 and 1.66 eV. Besides the movement of the absorption edge, the absolute values of the curves are found to diminish for higher Ga content. As previously mentioned in Chap. 2.4.2, an EQE curve describes the collection efficiency of the photo-generated charge carriers. Therefore, the overall suppression of the curve indicates higher loss in the device. Since all the layers except the CIGS absorber layer are identically fabricated, the higher loss originates most possibly from the increasing charge carrier recombination in the bulk of the CIGS absorber, or at the interfaces between buffer/CIGS or CIGS/Mo. In a working CIGS solar cell, the free holes moving in the valence band of the CIGS absorber E_V(CIGS) are transported through the CIGS/Mo-interface into the Mo electrode. As E_V(CIGS) stays energetically relative constant with Ga addition [41], the electrical property of the CIGS/Mo-interface should not significantly differ in solar cells with varying Ga content. Thus, if the increased recombination arises from interfaces, the most possible one is the CdS/CIGS-interface. This recombination issue will be discussed in more detail combining the results from jV, EQE and KPFM measurements in Chap. 5.4.

5.3 Fermi energy shifting in CuIn$_{1-x}$Ga$_x$Se$_2$ absorber layers

Figure 5.3 shows the KPFM measurement results on an untreated cross section of the sample with a GGI-ratio of 0.32. In the topography image the ZnO and CIGS layers can be easily distinguished on basis of different grain sizes of these two materials. In the CPD image the contrast between these two layers is even more distinct based on the fact that different materials possess different work functions. With the line section displayed in Fig. 5.3(c) the potential distribution through the CIGS/CdS/i-ZnO/ZnO:Al-

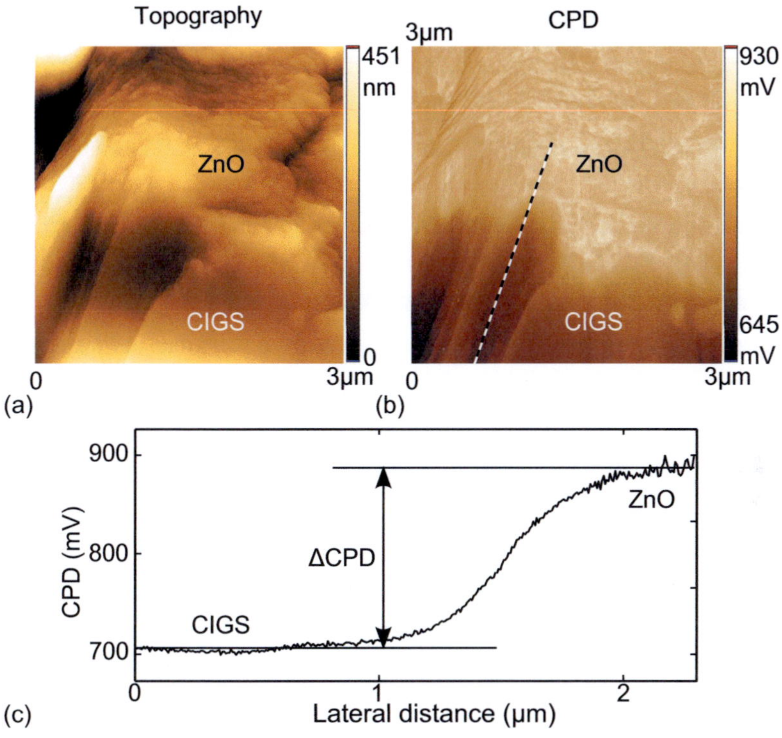

Figure 5.3: KPFM measurement results on an untreated cross section of the sample with a GGI-ratio of 0.32. The topography and CPD images are shown in **(a)** and **(b)**. In both images the ZnO and CIGS layers can be well distinguished. The line section marked with a dashed line in (b) is displayed in **(c)**. The CPD difference between the ZnO and CIGS layers is labeled as ΔCPD.

heterojunction can be comprehensively studied. This line section is extracted from the dashed line in the CPD image shown in Fig. 5.3 (b). The shape of this CPD curve follows the typical potential drop along a heteropn-junction formed by CIGS and ZnO. The difference between the CPD values on CIGS and ZnO is labeled as ΔCPD. Note that due to the tilted sample alignment the surface of the ZnO layer is also partially scanned, which can be verified by the dramatic reduction in the upper area of the topography image. Consequently, the ZnO layer seems to be thicker than the desired value of 400 nm, which is the sum of the thicknesses of the i-ZnO and ZnO:Al layers. However, this issue obviously has no influence on the analysis of ΔCPD.

In this way the ΔCPD values are recorded for all four samples with varying GGI-ratios. Practically, the CPD values will be influenced by the employed cantilevers [205, 206]. In order to minimize this influence, each sample is measured by five cantilevers in total. The measurement data are presented with different symbols in Fig. 5.4 (a). Despite of the offsets between the five cantilevers, an obvious tendency marked with the blue dashed line can be observed: ΔCPD increases up to a GGI value of 0.63 and drops steeply for the sample with no In content.

As reported by Wei $et~al.$ [41] in a simulation work, with an increasing GGI-ratio the CIGS valence band energy E_V(CIGS) decreases only slightly. Hence, the increase of the CIGS band gap energy E_g(CIGS) results almost exclusively in the increase of the CIGS conduction band energy E_C(CIGS). Based on this approximation, a qualitative energy diagram of the CIGS material can be illustrated as a function of the GGI value (see Fig. 5.4 (b)). Further quantities in the diagram include the vacuum energy E_{vac}, the Fermi energies of the ZnO and CIGS layers E_F(ZnO) and E_F(CIGS), and the work functions of the ZnO and CIGS layers ϕ(ZnO) $= E_F$(ZnO) $- E_{vac}$ and ϕ(CIGS) $= E_F$(CIGS) $- E_{vac}$. Since the ZnO layer in all solar cell devices are identically fabricated, E_F(ZnO) is depicted with a dashed line in the diagram as a constant reference level. ΔCPD is defined as the potential difference between ZnO and CIGS and is proportional to ϕ(ZnO) $- \phi$(CIGS). As E_F(ZnO) is constant, the variation of ΔCPD reflects indeed the variation of E_F(CIGS) as indicated by the blue dotted line. Note that the work function is defined for electrons with negative charges [153], the variation of E_F(CIGS) exactly mirrors the variation of ΔCPD. This outcome shows a clear Fermi energy shifting in CIGS absorbers with varying Ga content.

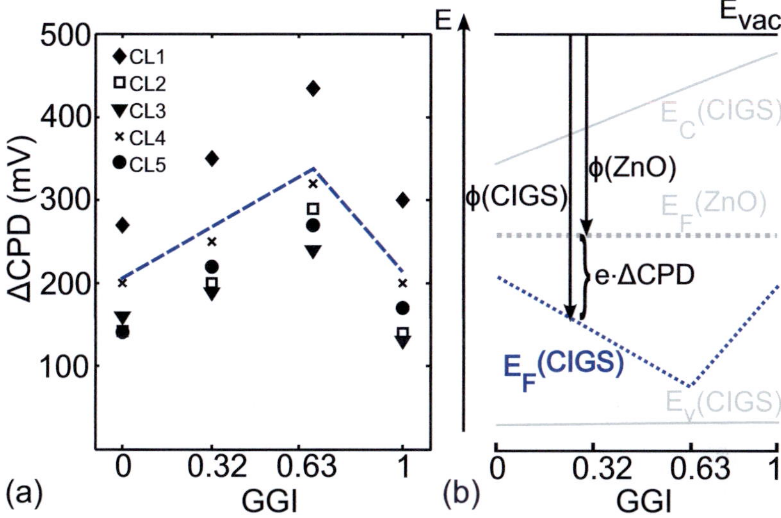

Figure 5.4: **(a)** Following the measurement principle introduced in Fig. 5.3(c) the ΔCPD values are recorded for all the four samples. Repeating this procedure with five different cantilevers (CL1 to CL5), the results reveal an offset between individual cantilevers. However, a clear tendency (blue dashed line) can be recognized for all the cantilevers. **(b)** The Fermi energy curve E_F(CIGS) (blue dotted line) derived by mirroring the blue dashed line in (a) shows a direct evidence for a Fermi energy shifting. The position of E_F(CIGS) to other energy levels is shown only qualitatively.

In more details, E_F(CIGS) approaches E_V(CIGS) for GGI-ratios between 0 and 0.63 and withdraws for GGI values from 0.63 to 1.

In general, the relative position of the Fermi energy over the valence band maximum reflects the density of free holes in p-type semiconductors like CIGS. In more detail, the smaller the energy distance the higher the density of free holes. In Ref. [41] it was calculated that the densities of single acceptors such as Cu vacancies (V_{Cu}), In or Ga vacancies (V_{III}) and Cu on In or Ga antisite (Cu_{III}) that can provide free holes are similar in both CuGaSe$_2$ and CuInSe$_2$. Only the acceptor levels in CuGaSe$_2$ are slightly shallower than that in CuInSe$_2$. However, the donor level Ga on Cu antisite (Ga_{Cu}) in CuGaSe$_2$ are energetically much deeper in the band gap than that of the In on Cu antisite (In_{Cu}) in CuInSe$_2$. The overall results show that, since there are more holes (shallower acceptors) and fewer compensating electrons (deep donors) in CuGaSe$_2$ as in CuInSe$_2$, the hole density in CuGaSe$_2$ is expected to be higher. Similar results were observed in an experimental work of Schröder *et al.* [207]. In that work solar cells based on epitaxially grown CIGS layers were investigated applying the temperature-dependent Hall measurement. By fitting the measurement data, two acceptor levels were found. By increasing the Ga content it came out that the acceptor density increases and the acceptor level depth decreases for both acceptor levels leading to a higher concentration of free holes.

Besides the Ga content, the Na content was also proposed to influence the p-type doping of the CIGS absorbers [208–211]. The Na content in the CIGS absorbers originate from the Na ions in the soda-lime glass substrates. During the deposition process of CIGS, the substrate is heated at nearly $600\,^\circ C$, which evokes the diffusion of Na ions into the CIGS absorber. Wei *et al.* [209] ascribed the major effect of Na to the elimination of the In on Cu antisite (In_{Cu}) donor defects, while Cahen *et al.* [208] and Kronik *et al.* [210] elucidated that the presence of Na evokes an oxidation-related passivation of the donor-like Se vacancies (V_{Se}). The occupation of V_{Se} by oxygen results in the formation of O_{Se}, which is a shallow acceptor. Indeed, the enhancement of the p-type conductivity of the CIGS layer was also experimentally proved (see Ref. [211]). Therefore, an analysis of the Na content in the CIGS films is made by means of sputtered neutral mass spectroscopy (SNMS) at ZSW. As shown by the SNMS results in Fig. 5.5 [29], the Na content increases systematically with the Ga addition. In Ref. [211] the authors have prompted that the presence of Na hinders the elemental interdiffusion of Ga. The higher the Na content in the CIGS layer, the less the preset Ga gradient is averaged. Interestingly, the finding in the current work shows that the Ga content has also an influence

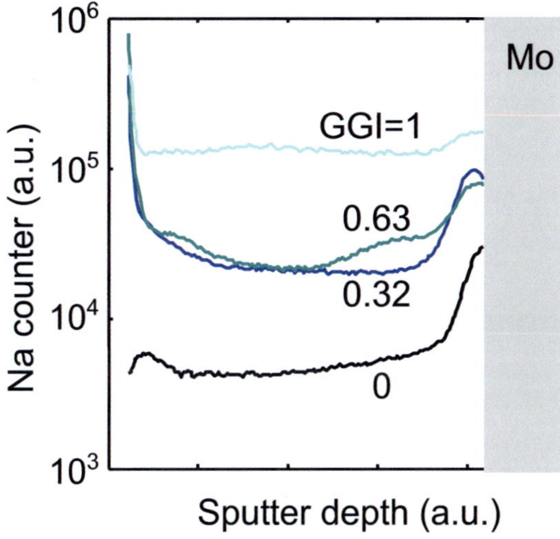

Figure 5.5: The Na content in CIGS absorbers with different GGI-ratios are investigated by Witte *et al.* [29] at ZSW using sputtered neutral mass spectroscopy (SNMS). The results show a systematic increase of the Na content with Ga addition.

on the Na diffusion and the increase of the Ga content is accompanied by the increase of the Na content. Since the increase of both Ga and Na will enhance the concentration of the free holes in the CIGS absorber, the observed Fermi energy shifting in the GGI range from 0 to 0.68 is most likely caused by both elements.

Schuler *et al.* [64] reported on the self-compensation of the intrinsic defects in CuGaSe$_2$ samples. Based on their temperature-dependent Hall measurements on a wide range of samples with different defect activation energies and defect densities, it was proposed that with increasing acceptor density the degree of compensation increases. This phenomenon was theoretically explained as follows: at a certain acceptor density if more acceptors are introduced, the Fermi level would shift down which lowers the formation enthalpies for donor defects even to negative values [212]. As a result, more donor defects will be formed increasing the compensation. In the energy band diagram, the Fermi energy will be shifted towards the middle of the band gap, which concurs with our results for GGI-ratios larger than 0.63. Furthermore, the slightly higher Cu content in the $CuGaSe_2$ sample (see Tab.5.1) may lead to less V_{Cu}, which is generally assumed as an acceptor state [41]. This may also contribute to the Fermi energy shifting towards the middle of the band gap.

5.4 Charge carrier recombination in $CuIn_{1-x}Ga_xSe_2$ solar cells

In Fig. 5.6 (a) an energy band diagram of the heterojunction of a CIGS solar cell in short circuit condition is sketched. Differently to the one in Fig. 5.4 (b), where a constant vacuum level E_{vac} in ZnO and CIGS layers was used, a constant Fermi energy E_F in both layers is applied. Physically, both configurations of the band diagram are identical. Thus, ΔCPD defined in Fig. 5.3 can be now interpreted as proportional to the variation of the local vacuum energy by a factor of the elemental charge e. Moreover, an important physical quantity the diffusion voltage V_D, which is also referred to as the built-in voltage in some literature, is indicated in the diagram. V_D is by definition the maximal potential drop in the conduction band through the junction in short circuit condition or in darkness [213]. Theoretically, V_D is the upper limit of the open circuit voltage V_{oc} [214]. According to the energy band diagram, the following relation exists between ΔCPD, V_D and three other quantities, i.e., the band gap energy of CIGS E_g(CIGS), the conduction band energy (electron affinity) of ZnO E_C(ZnO) and the valence band energy of CIGS E_V(CIGS)

$$e \cdot \Delta\text{CPD} + E_C(\text{ZnO}) = E_V(\text{CIGS}) - E_g(\text{CIGS}) + e \cdot V_D \qquad (5.1)$$

Thus, $e \cdot V_D$ can be described as

$$e \cdot V_D = e \cdot \Delta\text{CPD} + E_C(\text{ZnO}) + E_g(\text{CIGS}) - E_V(\text{CIGS}) \qquad (5.2)$$

With values of ΔCPD and E_g(CIGS) from the KPFM and EQE measurements and two additional literature values E_C(ZnO) and E_V(CIGS) acquired from the literature, $e \cdot V_D$ can be derived. In Fig. 5.6 (b) ΔCPD values are shown by blue diamonds with error bars containing all measurement points in Fig. 5.4 (a). E_g(CIGS) extracted in Fig. 5.2 (b) are depicted with black squares. In the literature, different values for E_C(ZnO) (e.g. 4.0 eV [215], 4.2 eV [216] and 4.6 eV [217]) and E_V(CIGS) (e.g. 4.64 eV [218], 5.0 – 5.2 eV [215]) were published. Theoretically, $e \cdot V_D$ is supposed to be between $e \cdot V_{oc}$ and E_g(CIGS) [214]. Therefore, the combination of E_C(ZnO) = 4.2 eV [216] and E_V(CIGS) = 4.64 eV [218] seems to be most reasonable for the current analysis. In this way, $e \cdot V_D$ is determined for each sample and the resulting data are shown in In Fig. 5.4 (b) by red diamonds. Values of V_{oc} determined in jV-measurements are presented in the same diagram.

Generally, the gap between V_D and V_{oc} reflects the loss mechanisms, particularly the recombination rate of the photogenerated charge carriers in a solar cell device [219]. The yellow area in Fig. 5.6 (b) shows clearly that the split-off between $e \cdot V_D$ and $e \cdot V_{oc}$ increases for higher Ga content. Thus, this finding concurs with the reduction of the absolute values of the EQE curves observed in Chap. 2.4.2 and can confirm the conclusions drawn in previous studies that the performance of CIGS solar cells with high GGI content is limited by stronger recombination processes. Note that the use of different values for E_C(ZnO) and E_V(CIGS) will lead to an integral movement of the curve $e \cdot V_D$ in the vertical direction in Fig. 5.6 (b). However, the form of the curve will not be changed. Therefore, the conclusion made above will not be influence. Furthermore, due the lack of the knowledge of the exact position of the recombination process based on the available techniques, and inconsistent opinions in the literature (dominating bulk recombination [42, 45, 46, 48, 49], dominating interface recombination [39, 47, 51, 56–58]), further investigations are needed for a complete understanding of this effect.

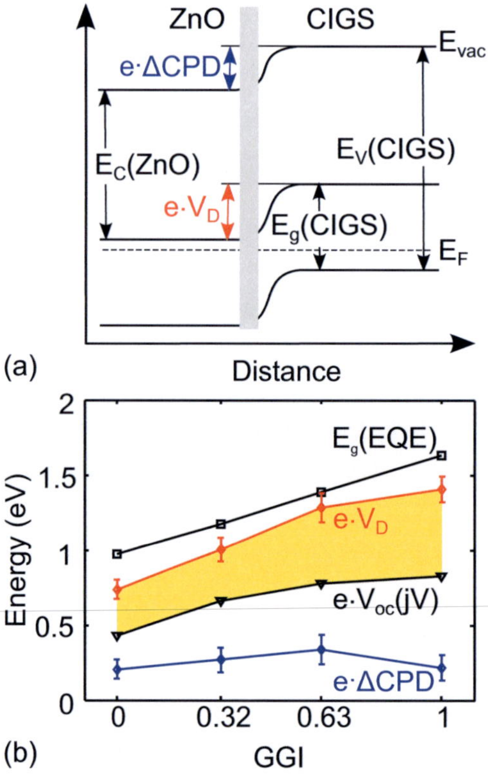

Figure 5.6: **(a)** A qualitative energy band diagram of the heterojunction of a CIGS solar cell in short circuit condition explaining the physical quantities ΔCPD, the diffusion voltage V_D, the ZnO conduction band energy $E_C(ZnO)$, the CIGS valence band energy $E_V(CIGS)$ and the CIGS band gap energy $E_g(CIGS)$. The relation between these five quantities is derived in Eq. 5.1. **(b)** With ΔCPD (blue diamonds) and $E_g(CIGS)$ (black squares) values from KPFM and EQE measurements and two additional literature values $E_C(ZnO) = 4.2\,eV$ [216], $E_V(CIGS) = 4.64\,eV$ [218], $e \cdot V_D$ (red diamonds) can be deduced from Eq. 5.2. For comparison, values of the open circuit voltage V_{oc} (black triangles) determined in jV-measurements are presented. The increasing gap between $e \cdot V_D$ and $e \cdot V_{oc}$ shown by a yellow area implies a higher recombination rate of free charge carriers in solar cells with higher Ga content.

Finally, it is necessary to briefly discuss the influence of the ambient conditions on the KPFM results. As it was introduced in Chap. 3, multiple factors may reduce the measured ΔCPD values from their true values. Indeed, taking all these effects into account, it can be estimated that the variation of ΔCPD will be more pronounced. As a result, the gap between $e \cdot V_D$ and $e \cdot V_{oc}$ would become even larger for high Ga content strengthening the main conclusion.

5.5 Chapter conclusion

CIGS solar cells with varying Ga content were investigated with macroscopic (jV and EQE) and microscopic (KPFM) measurements. The KPFM results showed a systematic shifting of the Fermi energy in the CIGS absorber material with Ga addition. The Fermi energy was observed to approach the valence band energy at GGI-ratios from 0 to 0.63, which is most likely due to the increase of both the Ga and Na content that enhance the p-type doping of the absorber. At GGI-ratios between 0.63 and 1 the Fermi energy was found to withdraw from the valence band energy, which originates possibly from a higher degree of the self-compensation due to the generation of compensating donor defects. Also, the higher Cu content in the sample without In content may contribute to the Fermi energy shifting away from the valence band energy. In addition, the results from jV, EQE and KPFM measurements were combined. The analysis indicated a higher recombination rate of free charge carriers in solar cells with higher Ga content, which can well explain the unsatisfying performance generally observed for these solar cells.

In this chapter, the CIGS solar cell with a GGI-ratio of 0.32 showed the highest power conversion efficiency. Hence, in the upcoming chapter CIGS solar cells with GGI-ratios around 0.3 will be carefully examined. In order to make the conclusions relevant for the industrial fabrication, these samples are produced in an inline multistage process that is comparable to the industrial standard.

6 Potential distributions at grain boundaries of CuIn$_{0.7}$Ga$_{0.3}$Se$_2$ absorbers

With CuIn$_{0.7}$Ga$_{0.3}$Se$_2$ absorbers thin-film solar cells with the highest power conversion efficiency were fabricated. As CIGS solar cells based on polycrystalline absorbers outperform their monocrystalline counterparts, the grain boundaries are widely considered to have a great importance on the performance of the solar cell. This chapter describes the analysis on the grain boundaries of a CuIn$_{0.7}$Ga$_{0.3}$Se$_2$ absorber layer. The potential variations at grain boundaries on the surface and on untreated cross sections of the absorber are investigated. A comparison between the results from the surface and cross sections evokes a discussion about the reevaluation of the conclusions drawn in previous studies.

6.1 Grain boundaries on the CuIn$_{0.7}$Ga$_{0.3}$Se$_2$ absorber surface

The CIGS absorber layer under investigation is fabricated by an inline multistage co-evaporation process. The Cu content (21.47 at.%) and the integral GGI-ratio (0.3) are determined by means of XRF at ZSW. The power conversion efficiency of the solar cell based on this absorber is 17%.

The measurements are at first performed on the surface of the CIGS absorber layer. As described in Chap. 4.2, the surface is exposed by selective etching using HCl. In Fig. 6.1 the measurement data in a dimension of $40 \times 40\,\mu\mathrm{m}^2$ are presented. In the topography image the grain structure is clearly resolved. In the CPD image different types of potential variations are observed at grain boundaries. According to their CPD line shape all grain boundaries can be classified as GB$_{dip}$ (dip-shape variation of the CPD

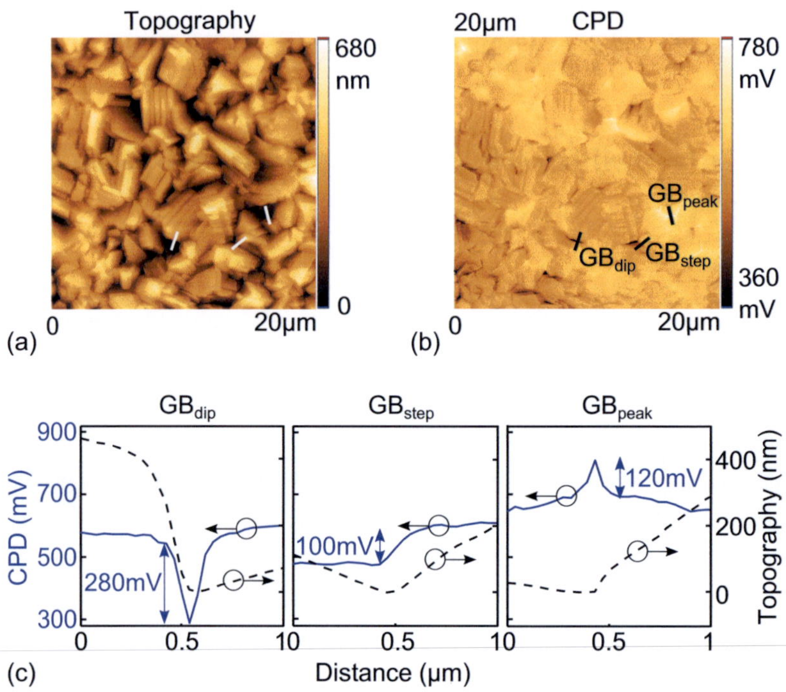

Figure 6.1: KPFM measurements on the surface of the $CuIn_{0.7}Ga_{0.3}Se_2$ absorber showing **(a)** topography and **(b)** CPD images. **(c)** CPD (blue solid lines) and topography (black dashed lines) line sections at three different grain boundaries marked in the images (a) and (b). According to the CPD line shape they are denoted as GB$_{dip}$, GB$_{step}$ and GB$_{peak}$.

signal), GB$_{step}$ (stepwise variation) and GB$_{peak}$ (peak-shape variation). It is obvious that there is no convolution between the topography and the CPD signal as shown by the line sections in Fig. 6.1 (c). The maximal magnitudes of CPD variations at these three types of grain boundaries are 280, 100, and 120 mV, respectively. These values match very well with former KPFM studies, which were also conducted in air [13, 145, 153], but they are slightly smaller than the values found under vacuum conditions [150]. This reduction can be attributed to the surface effects present in air, which were analyzed in Chap. 3.3.

In contrast to other studies, where only one type [13,14,74,145,153] or two types [150] of grain boundaries were reported, three types on the same sample surface are observed in the current study. This observation is consistent with a recent study conducted by Baier *et al.* [75], who investigated the surface of a $CuInSe_2$ sample. The inconsistency between different studies should be caused by multiple reasons, e.g., samples from different fabrication processes were studied, the measurements were carried out under different measurement conditions, or measurement data from different scanning ranges were analyzed. Obviously, the first two reasons wil noticeably influence the measurement results. However, the third reason was often neglected. Given the polycrystalline structure of the absorber layer, it can be easily understood that each type of grain boundary appears with a certain probability. Therefore, it is not adequate to draw conclusions based on the observation of only few grain boundaries. Instead, it should be based on observing larger areas, where sufficient grain boundaries of different types will emerge.

6.2 Grain boundaries on the $CuIn_{0.7}Ga_{0.3}Se_2$ absorber cross section

Figure 6.2 shows the topography and CPD images on an untreated cross section of the solar cell fabricated with the same CIGS absorber. The ZnO, CIGS, and Mo layers can be well distinguished in both images. Due to the tilted sample alignment, two phenomena can be observed in the CPD image. First, as already mentioned in the last chapter, the thickness of the ZnO layer appears to be enlarged, because the surface of the ZnO layer is also partly scanned. Second, the thickness of the CIGS absorber layer seems to be slightly reduced, because the measured thickness ($\approx 2\,\mu m$) is the projection of the real thickness ($\approx 2.2\,\mu m$) into the horizontal plane (see the sample alignment sketched in Fig. 4.6 (b)). Very importantly, in order

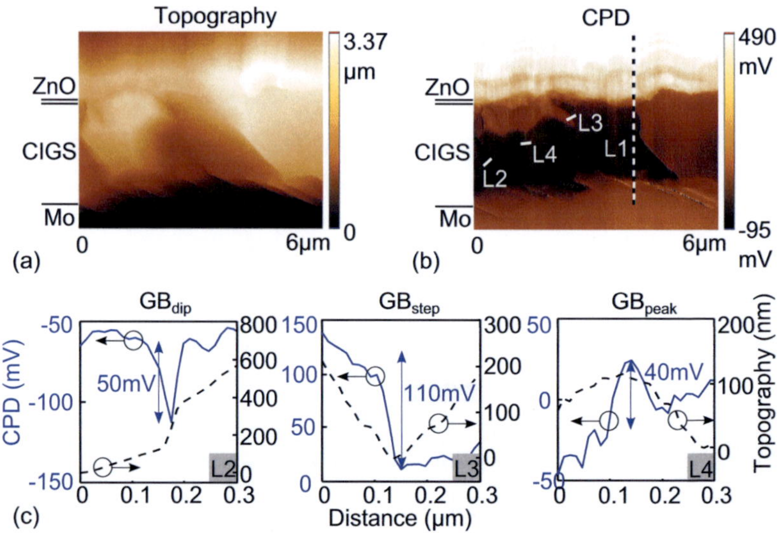

Figure 6.2: KPFM measurements on an untreated cross section of the solar cell based on the same $CuIn_{0.7}Ga_{0.3}Se_2$ absorber. The **(a)** topography and **(b)** CPD images with the size of $6 \times 4.8\,\mu m^2$ and a spatial resolution of $11.7\,nm/pixel$ are presented. In both images the ZnO, CIGS, Mo layers can be easily distinguished from top to bottom. The expected position of the CdS layer is indicated on the left side of the images with a double line. The line section L1 is analyzed in Fig. 6.3. **(c)** CPD (blue solid lines) and topography (black dashed lines) line sections at L2, L3, L4 indicated in (b), corresponding to the three types of grain boundaries GB_{dip}, GB_{step} and GB_{peak}, which are also observed on the absorber surface. However, the magnitude of potential variations at the GB_{dip} and GB_{peak}-type grain boundaries on the cross section are significantly smaller compared to the absorber surface.

to make the measurements on the cross sections comparable with the ones on the surface, all KPFM measurements are performed with the same scan parameters: the free oscillation amplitude of $17\,nm$, the set point of $13\,nm$ and the ac-voltage at the second resonance frequency $V_{ac}(f_1)=2\,V$ (details see Chap. 4.3).

Figure 6.3 displays the topography and CPD signals at the line section L1.

The position of L1 is marked with a dashed line in Fig. 6.2 (b). L1 exhibits the potential distribution through all layers of the solar cell. Again, both signals are obviously not convoluted. Note that the work function is defined for negatively charged electrons. Consequently, the CPD distribution corresponds to the inversed work function distribution [153]. Therefore, the CPD curve with a reversed coordinate adopted in Figure 6.3 (a) follows the curve shape of the work function ϕ. The work function ϕ is defined as the energy difference between the vacuum energy E_{vac} and the Fermi

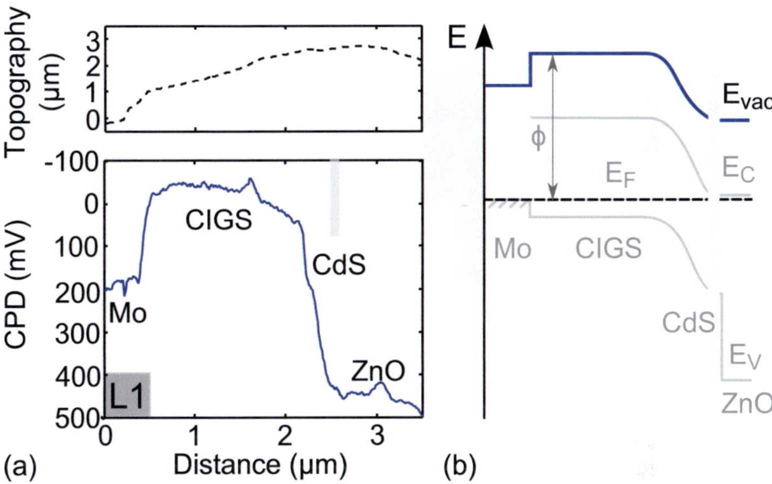

Figure 6.3: **(a)** Topography and CPD line profiles taken at the position L1 in Fig. 6.2(b). This line sections covers the whole Mo/CIGS/CdS/ZnO-structure. The potential variation caused by CdS (gray bar) is difficult to analyze, because this very thin layer is frequently covered by the ZnO overlayer after cleavage. Since the work function ϕ is defined for negatively charged electrons, the CPD distribution mirrors the ϕ distribution. Therefore, for a better comparison with the energy diagram in (b), an inverted coordinate for the CPD values is used. **(b)** A qualitative band diagram including all the layers in the solar cell. As the sample is short circuited, Fermi energies in all the layers are at the same level. Therefore, the variation of the work function $\phi = E_{vac} - E_F$ reflects in the variation of the vacuum level E_{vac}.

energy E_F. Since the sample is short circuited, the Fermi energies E_F in different materials are at the same level as depicted in the qualitative energy band diagram in Fig. 6.3 (b). With this configuration the variation of ϕ reflects in the variation of the local E_{vac} as indicated by the thick blue line in the energy diagram. In this line section the extracted potential difference between CIGS and ZnO is $450 - 500\,mV$, which is a value comparable to those of $550 - 600\,mV$ observed under UHV conditions [182]. This outcome indicates the cleanliness of the cleaved cross sections demonstrated in this work, which plays a decisive role in the interpretation of the observed results.

It is worth noting that, since the very thin CdS buffer layer is frequently covered by the ZnO overlayer after cleavage, it is difficult to analyze the potential fluctuation caused by this layer. Nevertheless, the position of this layer can be still roughly determined in the following way. Due to the high doping concentration of the ZnO:Al layer, the space charge region in this layer is negligibly small. Hence, the variation of the bands can be considered to end at the i-ZnO/ZnO:Al-interface, which corresponds to the right edge of the gray bar in Fig. 6.3 (a). With that the CIGS/CdS-interface can be further determined at $100\,nm$ (sum of the thicknesses of the CdS and i-ZnO layers) from the i-ZnO/ZnO:Al-interface towards the CIGS layer. Thus, the position of the CdS layer should be located within the gray bar.

The topography and CPD line sections of L2, L3, and L4, which are marked in Fig. 6.2 (b), are illustrated in Fig. 6.2 (c). They show the three types of grain boundaries GB_{dip}, GB_{step}, and GB_{peak}, which were already found on the surface. In agreement with the observation on the surface, grain boundaries of the type GB_{dip} and GB_{step} appear more frequently than the GB_{peak}-type ones on cross sections. However, very importantly, the height of the CPD variations of the types GB_{dip} ($50\,mV$) and GB_{peak} ($40\,mV$) on the cross section are much smaller than on the surface.

Because the sample preparation and the KPFM measurements were conducted in air, the cleavage process of the sample will lead to surface oxidation and atomic reconstruction at the freshly produced surface. However, benefiting from the abandonment of the polishing and the subsequent cleaning processes, a simple cleavage still maintains the cross section in a nearly unmodified condition. Consequently, the properties of grain boundaries on the untreated cross sections will be very close to those of the buried GBs (at least more closely as the ones on the absorber surface). Therefore, a comparison between the results from the surface and the cross section of the same CIGS absorber layer leads to the following two conclusions.

First, the GB_{dip}-type is dominating in quantity. This means that the work function ϕ increases at most grain boundaries. According to Refs. [75, 150], this type of work function variation indicates negative charges at grain boundaries. If the photogenerated charge carriers move across GB_{dip}, the electrons will be repelled while the holes will be attracted into the grain boundaries. However, since the value of GB_{dip} is much smaller than that on the absorber surface, a contingent profit provided by the grain structure in terms of the charge carrier separation will be much weaker compared to the estimations of previous studies [13, 14, 145, 150]. Thus, the mystery that solar cells based on polycrystalline CIGS absorbers outperform those based on monocrystalline absorbers in efficiency should be explained rather by other beneficial effects. The possible ones are the formation of the OVC layer on the absorber surface, which enables a homojunction and thus may effectively reduce the charge carrier recombination at the interface [185] or the formation of a band grading that works as a back surface field and may prevent the charge carrier recombination at the back contact [204]. Taking advantage of more flexibility in the fabrication, these effects can be more easily acquired by manufacturing polycrystalline CIGS absorber layers.

Second, stepwise potential variations at the GB_{step}-type in the order of $100\,mV$ are observed. Previously, such step-shaped potential variations have been observed between different facets of single grains [220] and at $\Sigma 3$ grain boundaries [73] on the surface of epitaxially grown $CuGaSe_2$ layers and recently also at $\Sigma 3$ grain boundaries on the surface of a polycrystalline

$CuInSe_2$ layer [75]. This phenomenon was attributed to different surface dipole characteristics for different crystal orientations [75, 220]. This type of potential variations arise mostly at charge free grain boundaries or facets, which are harmless for the charge carrier recombination [75].

Consequently, the observation in the current study shows that grain boundaries in CIGS absorbers are either slightly charged ones (mostly GB_{dip}, only small amount GB_{peak}) with small potential variations or charge neutral ones with comparatively larger potential variations. This finding suggests the relatively inactive properties of grain boundaries in CIGS absorber layers agreeing with the conclusions drawn in the outstanding review paper published by Rau *et al.* [12]. It provides a reasonable explanation for the superior performance of CIGS solar cells despite the abundance of grain boundaries in the absorber, which are commonly regarded as detrimental for the performance of a semiconductor device.

One crucial issue remains to be discussed, namely the identification of grain boundaries. In general, grain boundaries are located with a very high possibility at positions where large topography variations emerge. Therefore, the topography images were used as the most important reference for finding grain boundaries in previous studies. However, at individual positions with a less distinct topography feature, the determination of grain boundaries without additional measurement techniques such as electron backscatter diffraction (EBSD) [75] could be erroneous. Nevertheless, this risk can be effectively reduced by observing sufficiently large areas, which can ensure the analysis of adequate grain boundaries. This requirement is easily fulfilled by multiple measurements on absorber surfaces with a scan area of $20 \times 20 \, \mu m^2$ for each (see additional data in Fig. 6.4). For cross section measurements an area with a length over $60 \, \mu m$ in the horizontal direction in total is under investigation. Some examples are exhibited in Fig. 6.5. Based on the analysis of more than 40 grain boundaries on cross sections, the occurrence probabilities of different types of grain boundaries are about 70% (GB_{dip}), 25% (GB_{step}), and 5% (GB_{peak}).

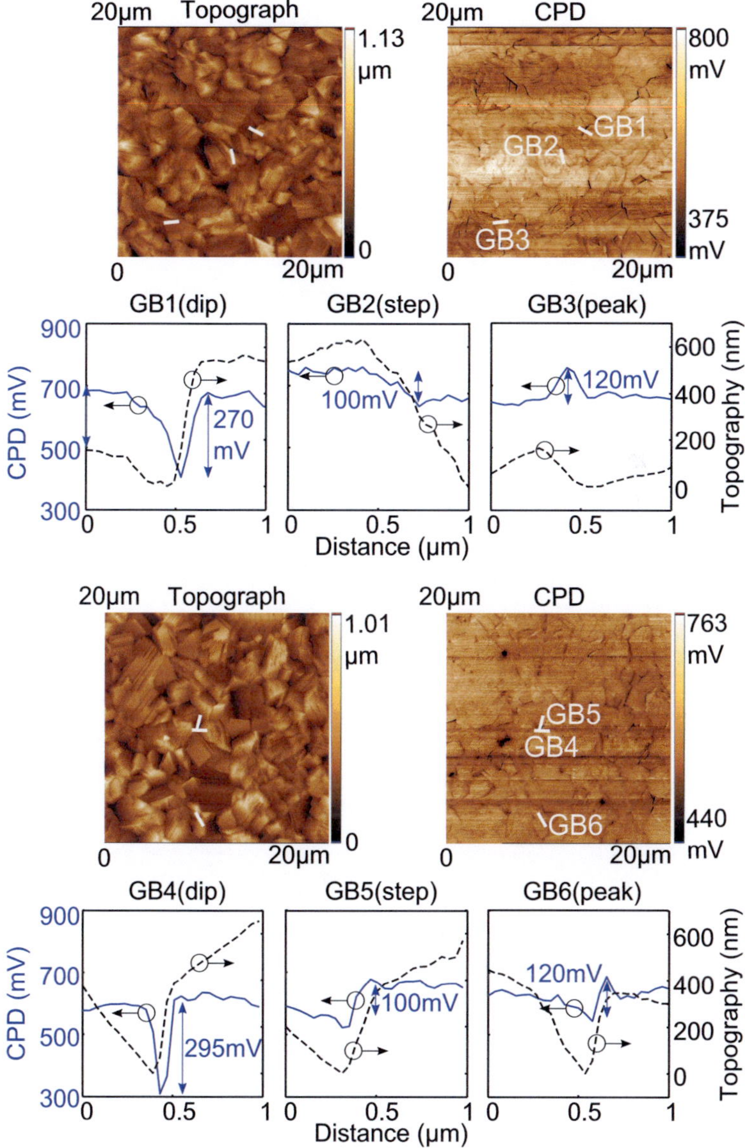

Figure 6.4: Additional KPFM data of potential variations at grain boundaries on surfaces of CIGS absorbers.

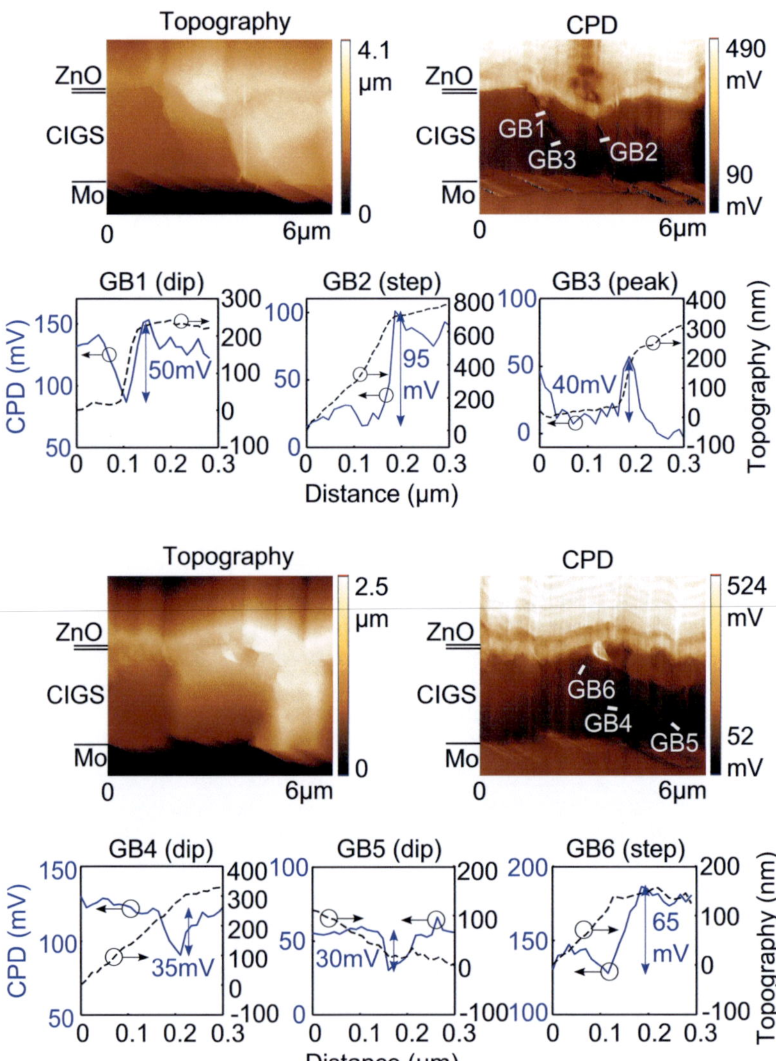

Figure 6.5: Additional KPFM data of potential variations at grain boundaries on cross sections of CIGS absorbers.

Moreover, some KPFM measurements on cross sections are so stable, that several consecutive measurements can be joined to form a panorama. In Fig. 6.6 such an image is demonstrated, which provides a good overview of the lateral potential fluctuation of the solar cell heterojunction. Basically, if a single CIGS grain with the overlayers on its top is considered as the heterojunction of a mini solar cell, the macroscopic solar cell is then a parallel circuit of a large number of these mini solar cells. Thus, the potential fluctuation in lateral direction may indicate a fluctuation of the open circuit voltages delivered by individual mini solar cells. Since the open circuit

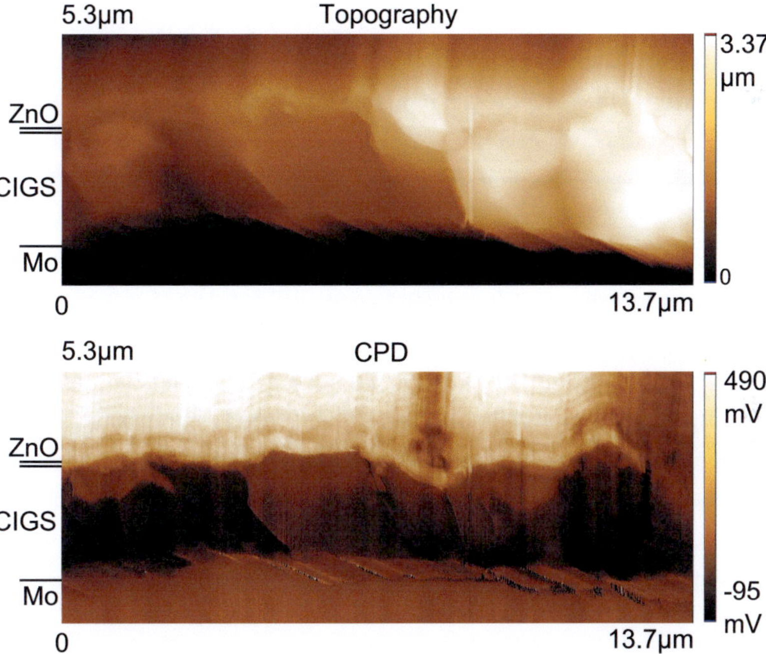

Figure 6.6: Panoramic images containing **(a)** topography and **(b)** CPD images are made by combining three consecutive measurements. The combined images have a size of $13.7 \times 5.3\,\mu\text{m}^2$ and a spatial resolution of 11.7 nm/pixel. With these images a better overview of the grain structure and the potential variations therein can be obtained.

voltage of the macroscopic solar cell will be limited by the lower values of the open circuit voltages of the mini solar cells, the current finding could be one of the critical issues that hinder CIGS approaching their theoretical limits [221].

6.3 Chapter conclusion

Potential variations at grain boundaries on the surface and cross sections of the same CuIn$_{0.7}$Ga$_{0.3}$Se$_2$ absorber layer were investigated. The results showed three types of grain boundaries depending on the CPD line shape on both the surface and cross section. The results on the absorber surface were comparable with the ones from former studies by other groups, whereas the magnitudes of the potential variations at grain boundaries on cross sections were found to be noticeably smaller than those on the surface. Since the properties of grain boundaries on cross sections are close to those buried in the bulk, it is important to evaluate the functionality of grain boundaries based on results from cross sections. Thus, the results showed that the beneficial functionality of CIGS grain boundaries in terms of charge carrier collection and transport should be much weaker than estimated before.

Up to now, the potential variations were observed in darkness. However, illuminating the solar cells has undoubtedly a great significance for understanding their real properties, since the performance of these devices is ultimately determined under illumination. Therefore, in the next chapter a white light illumination will be inserted during the scanning, in order to simulate the standard test conditions for solar cells.

7 Potential distributions in $CuIn_{0.7}Ga_{0.3}Se_2$ solar cells under illumination

In the last chapter grain boundaries in a $CuIn_{0.7}Ga_{0.3}Se_2$ absorber were studied in dark condition. In this chapter the solar cell based on the same absorber is investigated under white light illumination. The gradual changing of the potential distribution through the solar cell heterojunction is observed at defined illumination intensities. Moreover, the potential variations at grain boundaries in the absorber are analyzed under an illumination intensity similar to the standard test conditions. The findings in darkness from the last chapter together with the outcome of this chapter lead to a general interpretation of the functionality of CIGS grain boundaries in working solar cells.

7.1 Solar cell heterojunction under defined illumination intensities

Following the same procedure the potential distributions on cross sections of the solar cell based on the $CuIn_{0.7}Ga_{0.3}Se_2$ absorber are investigated in darkness again. The results achieved from three cross sections of the solar cell are presented in Fig. 7.3 (a), (c) and Fig. 7.4 (a). In topography images the ZnO, CIGS and Mo layers can be well distinguished as marked on the side. In CPD images a clear contrast between the layers can be observed. A CPD line section through the ZnO/CdS/CIGS-heterojunction is extracted from the position marked by a dashed line in Fig. 7.4 (a). The data shown at the bottom of Fig. 7.1. As indicated by the double arrow, the potential drop through the heterojunction , which was already defined in Chap. 5 as ΔCPD,

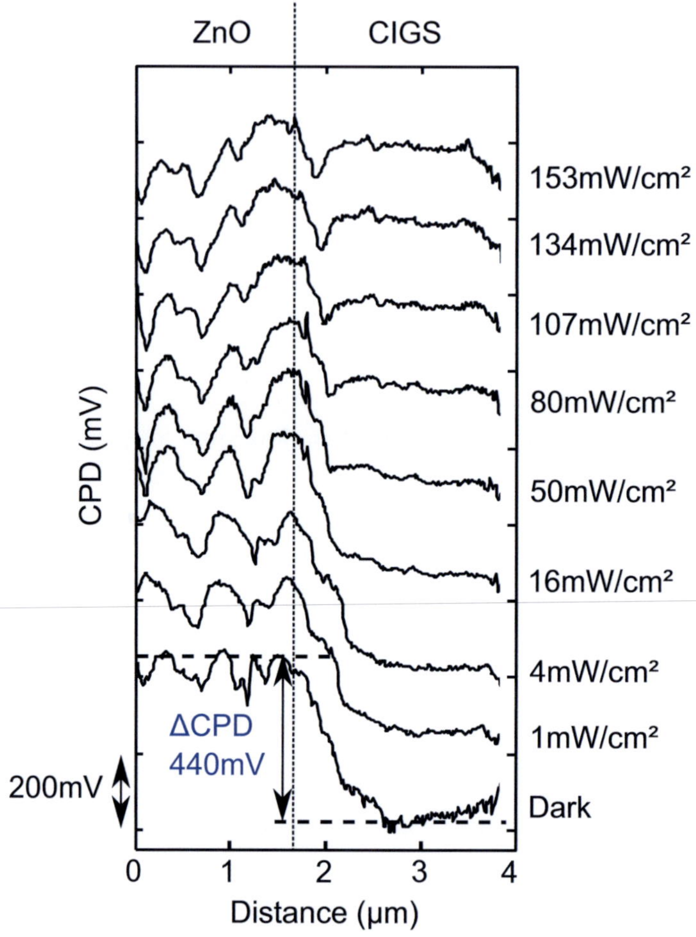

Figure 7.1: From bottom to top, the potential distribution through the ZnO/CdS/CIGS-heterojunction in darkness and under increasing illumination intensities. The potential difference between ZnO and CIGS is denoted as ΔCPD and its value in darkness is 440 mV. With increasing illumination intensity ΔCPD decreases and vanishes at intensities beyond 107 mV/cm^2. The line sections are extracted at the dashed line in Fig. 7.4.

has a value of 440 mV in darkness. Subsequently, the solar cell is illuminated through the ZnO top electrode with gradually increased intensity and the CPD distributions at the same position are recorded. The measurement data and the corresponding illumination intensities are shown in Fig. 7.1. It can be clearly seen that ΔCPD decreases with increasing illumination intensity and flat-band conditions (flattening of the potential drop) are achieved at intensities beyond 107 mW/cm^2. This is an intensity similar to the standard test conditions for solar cells (100 mW/cm^2, AM 1.5).

For clarity, ΔCPD values at all illumination intensities are extracted and shown in Fig. 7.2 with blue triangles. Simultaneously to the KPFM measurements, the other two quantities, i.e., the photovoltage V_{Photo} and the photocurrent I_{Photo} are recorded between the electrodes of the cleaved solar cell. Basically, the photovoltage V_{Photo} and the photocurrent I_{Photo} are the open circuit voltage V_{oc} and short circuit current I_{sc} at a given illumination level. Their values are depicted in Fig. 7.2 with red circles and black squares, respectively. It can be clearly observed that with increasing illumination intensity V_{Photo} increases and saturates at 667 mV and I_{Photo} increases almost linearly. Note that the open circuit voltage V_{oc} of the initial solar cell before cleavage under standard test conditions was 671 mV, which is almost identical to the maximal value of V_{Photo}. This together with the linear increase of I_{Photo} shows that the cleaved solar cell is still fully functional.

In Chap. 2.3.1 it was introduced that under illumination the Fermi energy E_F splits into the quasi Fermi energies E_{Fn} and E_{Fp} in the n-type ZnO and p-type CIGS, respectively. This results in a emergence of the photovoltage $V_{Photo} = (E_{Fn} - E_{Fp})/e$ between the junction edges, where e stands for the elemental charge. Meanwhile, the potential drop through the pn-junction reduces from the initial diffusion voltage V_D in darkness to $V_D - V_{Photo}$. Ideally, if there were no surface states and surface dipoles that typically cause energy band bending or band offsets, the variation of ΔCPD acquired on the surface would be identical to that of V_D in the bulk material. In real-

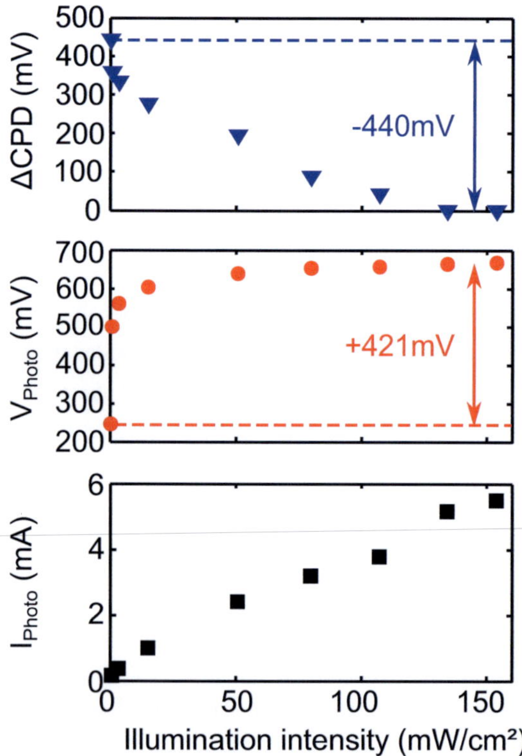

Figure 7.2: ΔCPD values (blue triangles) are determined from Fig. 7.1 for each illumination step. Simultaneously to the KPFM measurements, the photovoltage V_{Photo} (red circles) and photocurrent I_{Photo} (black squares) are measured between the electrodes of the cleaved solar cell in the open circuit and short circuit condition, respectively. The reduction of ΔCPD of 440 mV (blue arrow) agrees well with the increase of V_{Photo} by 421 mV (red arrow). Moreover, I_{Photo} increases linearly with the illumination intensity.

ity, if the material surface is not severely manipulated by surface states and dipoles, the behaviors of ΔCPD and V_D are still approximately the same. Therefore, the relation between ΔCPD under illumination and in darkness will be $\Delta\text{CPD}_{illumination} = \Delta\text{CPD}_{darkness} - V_{Photo}$.

The results in Fig. 7.2 show a reduction of ΔCPD by 440 mV (blue arrow) and an increase of V_{Photo} by 421 mV (red arrow), which agrees very well with the just mentioned relation for ΔCPD. This reveals that the untreated cross sections are not dominated by strong surface effects. To be pointed out, as mentioned in Chap. 3.3.3, V_{Photo} in darkness has a non-zero value of 246 mV. This results from the larger diameter of the laser beam than the width of the cantilever, which leads to a exposure of the surrounding area of the scan point in the laser light. Even if the point just being scanned may be not directly illuminated by the laser, the photogenerated charge carriers in the surrounding area can diffuse to this position and generate a photo-voltage. Limited by this technical issue, it is impossible to observe the potential distribution in a real dark condition with the current experimental setup. Nevertheless, a simple addition of ΔCPD and V_{Photo} that are acquired without deliberate illumination provides a good estimation of the diffusion voltage $V_D = 440\,\text{mV} + 246\,\text{mV} = 686\,\text{mV}$ in real darkness. This value is larger than the open circuit voltage V_{oc}, which is compatible with the fact that V_D in equilibrium (junction short circuited or in darkness) is the theo-retical upper limit of V_{oc} [214]. Certainly, since the sample preparation and KPFM measurements were carried out in air, the influence of the ambient conditions on the reduction of the potential contrast between materials can be never totally excluded. Thus, the real value of V_D is supposed to be even higher.

7.2 Grain boundaries under white light illumination

In the last chapter three types of potential variations at grain boundaries GB_{dip}, GB_{step} and GB_{peak} were observed on untreated cross sections of

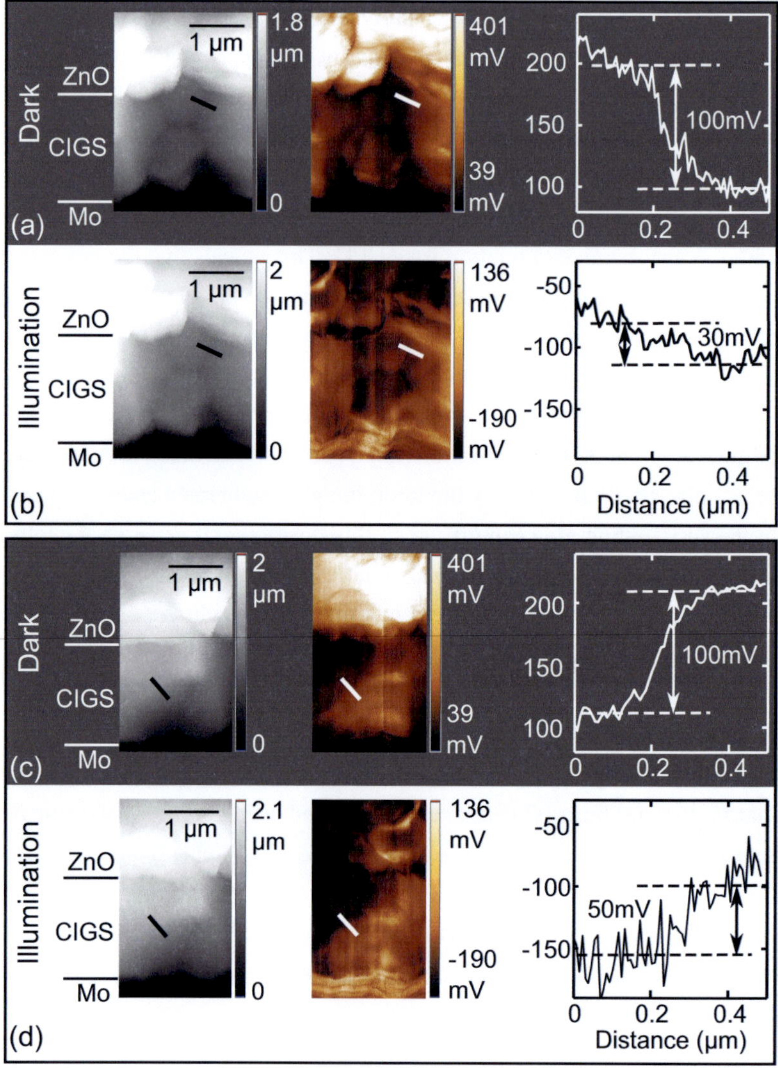

Figure 7.3: KPFM measurements on two untreated cross sections in darkness (dark background) and under an illumination intensity of 134 mW/cm^2 (white background). The topography images, CPD images and CPD line sections for the step-shaped potential variations at positions marked with solid bars are presented in the first, second and third column, respectively.

CIGS absorbers in darkness. Among them GB_{step} exhibited the largest magnitude of 100 mV. In Fig. 7.3 (a), (c) and Fig. 7.4 (a) potential steps at GB_{step} in darkness from three cross sections are shown. Their values around 100 mV are consistent to the former observations. As it was discussed in Chap. 6.2, despite of the electrically inactive properties of this type of potential variation, the potential fluctuations between the CIGS grains can still negatively affect the open circuit voltage V_{oc} and consequently the solar cell performance. In order to evaluate the real impact of the CIGS grain structure on operating solar cells, the behavior of the potential variations at grain boundaries, especially that of GB_{step}, is investigated under illumination.

In Fig. 7.3 (b), (d) and Fig. 7.4 (b) the potential distributions on the same cross sections at an illumination intensity of $134 \, mV/cm^2$ are imaged. The potential contrast between the layers nearly vanishes and the potential steps are reduced down to less than half of their initial values in darkness. At the two positions in Fig. 7.3 even no clear steps can be recognized. Physically, this finding can be explained with the screening effect discussed in Chap. 3.3.1. In more detail, the photogenerated charge carriers can screen the local surface state charge and thereby flattening the surface band bending. As a result, the difference between the surface band bending at facets or grains is reduced under illumination. Moreover, potential variations at the other two types of grain boundaries GB_{dip} and GB_{peak} are found to reduce to even less than 20 mV, which makes them hardly observable in the CPD images. This means that the largest potential variations within the CIGS polycrystalline structure in operation are only in the order of 50 mV. In comparison to the diffusion voltage induced by the solar cell heterostructure, potential variations at grain boundaries in such an order are most likely not enough to provide a remarkable beneficial effect in terms of collection and transport of the free charge carriers. This finding can support our conclusion made in the last chapter that CIGS grain boundaries have rather inactive properties. This feature is assumedly one of the most decisive factors for the superior performance of CIGS solar cells among all kinds of polycrystalline solar cells.

7.3 Influence of illumination on surface conditions

Since KPFM is a highly surface-sensitive measurement technique, the surface conditions of the sample are crucial to the measurement results and their final interpretation. As introduced in Chap. 3.3.2, the surface conditions can be easily influenced by a lot of factors, e.g., the illumination during the measurement. Therefore, KPFM measurements are performed at the same position in a sequence of in darkness, under illumination and in darkness again. All measurement data are shown in Fig. 7.4. In the CPD images in darkness for the second time the potential contrast between ZnO and CIGS and the potential steps at GB_{step} are reduced from 440 mV to 300 mV and 100 mV to 80 mV, respectively. Note that V_{oc} and I_{sc} of the cleaved solar cells do not show observable degradation for a time period as long as several months, the reduction of the CPD contrast should be attributed to surface effects such as long-lived deep trap states and surface aging. Long-lived deep trap states were previously reported on organic solar cells [222]. Due to their existence the potential contrast was found to fully recover until two days later. In this study, the potential contrast does not fully recover to the original value, even if the sample is kept in darkness for days. Indeed, it gradually reduces over time. Therefore, the observed reduction of the potential contrast after illumination is more likely due to the surface aging. Even if the cross sections are not directly illuminated, the heat generated in the solar cell may accelerate aging processes such as oxidation or atomic reconstruction on the surface of the cross section. However, this slight aging effect should not affect the conclusions made above.

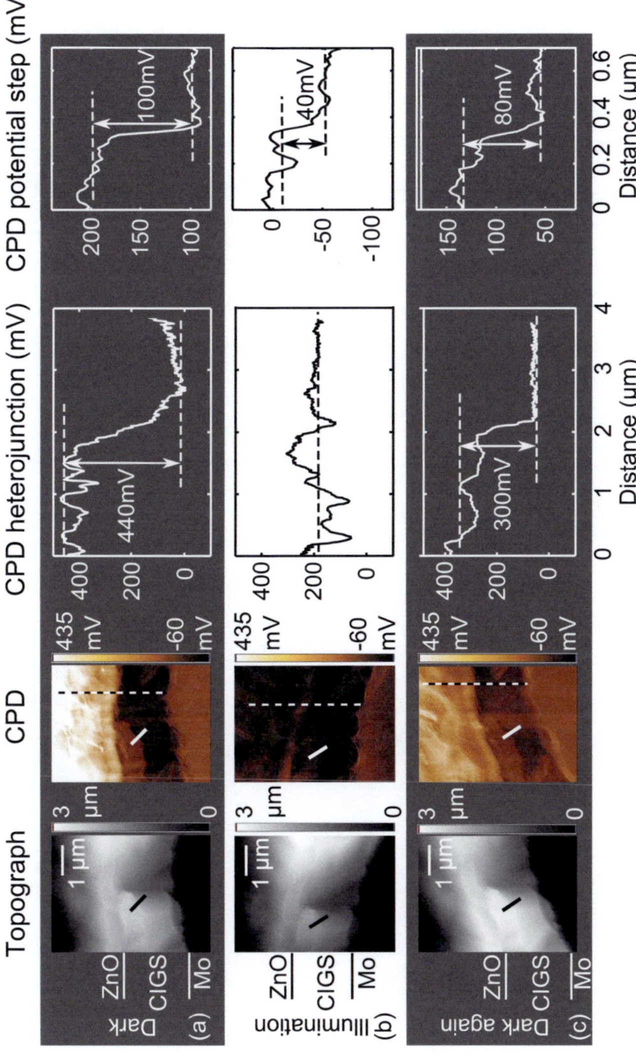

Figure 7.4: In order to find out the influence of the illumination on the surface conditions, KPFM measurements are performed in darkness (dark background), under illumination with an intensity of 134 mV/cm^2 (white background) and then again in darkness (dark background). After the illumination the values of the potential drop through the heterojunction (third column, extracted from dashed lines) and the potential steps at GB$_{step}$ in the CIGS absorber (fourth column, extracted from solid bars) are reduced from 440 mV to 300 mV and 100 mV to 80 mV, respectively.

7.4 Chapter conclusion

Potential distributions on cleaved cross sections of the thin-film solar cell based on a $CuIn_{0.7}Ga_{0.3}Se_2$ absorber were investigated under controlled illumination intensities. The potential drop through the solar cell heterojunction was observed to decrease with increasing illumination intensity and the magnitude of its decrease correlates well to the increase of the photovoltage. In addition, the short circuit current was found to increase linearly with the illumination intensity. These findings indicated that the cleaved solar cell is still fully functional and its untreated cross sections are not strongly manipulated by surfaces states or dipoles. The largest potential variations in the CIGS grain structure was found to reduce from $100\,mV$ in darkness down to less than $50\,mV$ under an illumination intensity comparable to the standard test conditions. Thus, the studies described in the last and this chapter show direct evidences for the inactive properties of the CIGS grain structure.

After the potential distributions in CIGS solar cells with the conventional layer stacking were thoroughly investigated in the last two chapters, another highly interesting topic will be studied in the upcoming chapter, i.e., why CIGS solar cells with a ZnS/ZnMgO buffer system underperform those with a conventional CdS/i-ZnO buffer system.

8 Comparison between ZnS/(Zn,Mg)O and CdS/i-ZnO buffer systems

In this chapter the performance of CIGS solar cells with the ZnS/(Zn,Mg)O alternative buffer system is compared with that of the solar cells using the conventional CdS/i-ZnO buffer system. Particularly, the potential distributions through the heterojunctions formed with these two types of buffer systems are carefully compared with KPFM, in order to find the reason for the generally observed loss in the open circuit voltage in CIGS solar cells with the ZnS/(Zn,Mg)O buffer system.

8.1 jV-characteristics

Figure 8.1 depicts the typical jV-curves of CIGS solar cells with ZnS/(Zn, Mg)O (blue solid line) and CdS/i-ZnO (green dashed line) buffer systems. For a better comparison, the significant photovoltaic parameters determined in the jV measurements are exhibited in Tab. 8.1. Note that except the buffer system all the other layers including Mo, CIGS and ZnO:Al are identically fabricated. Moreover, the jV-curve and the corresponding parameters of the solar cell with ZnS/(Zn,Mg)O buffer system are acquired after a light soak of 30 minutes at $200\,^{\circ}C$. As reported by Witte *et al.* [79], these solar cells typically show transient effects and such a treatment can noticeably improve the fill factor FF and the open circuit voltage V_{oc}.

It comes out from the comparison that the solar cell with the ZnS/(Zn,Mg)O buffer system delivers a higher short circuit current density j_{sc}. This is due to the larger band gap energy of ZnS, which results in less absorption loss in the blue wavelength region. However, the gain in j_{sc} is accompanied

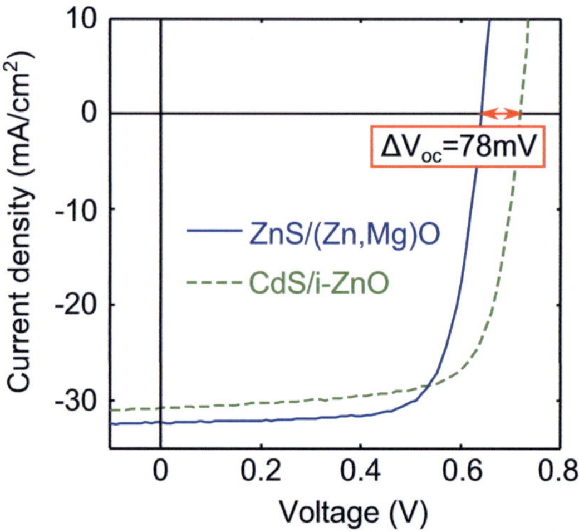

Figure 8.1: jV-curves of CIGS solar cells with ZnS/(Zn,Mg)O (blue solid line) and CdS/i-ZnO (green dashed line) buffer systems. The solar cell with the ZnS/(Zn,Mg)O buffer system delivers a larger short circuit current density, however, a reduced open circuit voltage V_{oc} by 78 mV. The jV-measurements are performed at ZSW.

Buffer system	j_{sc} (mA/cm^2)	V_{oc} (mV)	FF (%)	η (%)
ZnS/(Zn,Mg)O	32.4	640	73.8	15.3
CdS/i-ZnO	30.9	718	72.5	16.1

Table 8.1: Significant photovoltaic parameters extracted from the jV-measurements in Fig.8.1.

by a decrease in the open circuit voltage V_{oc}, which limits the solar cells reaching a higher efficiency. The loss in V_{oc} is generally observed in CIGS solar cells with the ZnS buffer layer [25, 30, 223] and also other alternative buffer materials [224, 225]. In the chosen samples the reduction of V_{oc} is 78 mV.

8.2 Potential distribution through the heterojunction of the solar cells

Figure 8.2 shows the topography and CPD images of CIGS solar cells fabricated with ZnS/(Zn,Mg)O and CdS/i-ZnO buffer systems. The Mo, CIGS and ZnO layers can be easily distinguished in the topography images. Due to the identical fabrication the morphology of these layers in both solar cells are quite similar. Also in the CPD images no distinct difference, e.g., abrupt potential peak or dip at the buffer system, can be observed. For a more accurate analysis, CPD line sections from these two samples (blue solid line for ZnS/(Zn,Mg)O and green dashed line for CdS/i-ZnO) are extracted and compared in Fig. 8.3.

Based on the CPD variation and also assisted by the topography variation, the CPD line sections can be separated in four parts: the surface of the ZnO layer, the cross section of the ZnO layer, the space charge region (SCR) in the CIGS layer and the neutral region in the CIGS layer. Therewith, the CPD line sections at heterojunctions formed with two types of buffer systems can be comprehensively compared and the results unveil two major differences. First, the CPD difference between the ZnO and CIGS layers, which was labeled previously as ΔCPD, in the ZnS/(Zn,Mg)O sample (400 mV) is about 100 mV smaller than that from a CdS/i-ZnO sample (500 mV). Second, the width of SCR in the CIGS layer of the ZnS/(Zn,Mg)O sample (1280 nm) is obviously smaller than that in the CdS/i-ZnO sample (890 nm). Since depending on the cantilever in use, variations in the CPD distributions may occur (see Chap. 5.3), the CPD line sections are extracted from cross sections measured with different cantilevers. Despite slight deviations in the absolute values of ΔCPD and the width of SCR, the comparison based on measurement data acquired with the same cantilever is always consistent. To be mentioned, slightly different to the CPD data presented in Chap. 6 and 7, the surface and the cross section of the ZnO layer reveal different CPD values. This effect is ob-

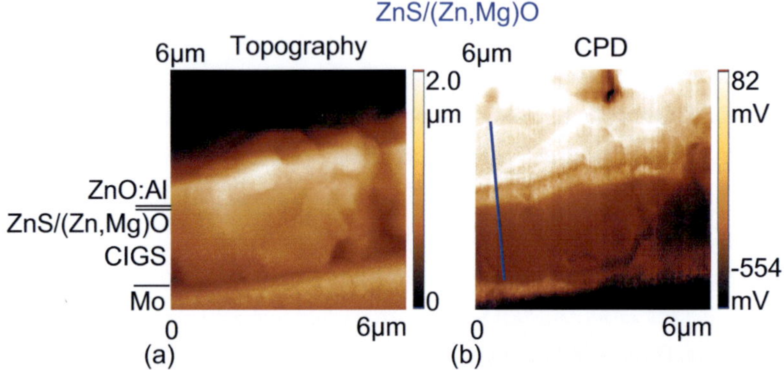

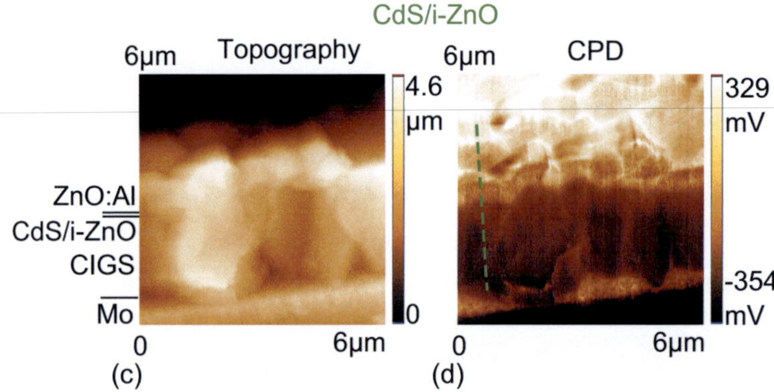

Figure 8.2: Topography and CPD images on cross sections of CIGS solar cells with ZnS/(Zn,Mg)O (**(a)** and **(b)**) and CdS/i-ZnO (**(c)** and **(d)**) buffer systems. No distinct difference can be recognized in the topography or CPD images from the two samples. For a closer analysis, CPD line sections from the sample with ZnS/(Zn,Mg)O (blue solid line) and CdS/i-ZnO (green dashed line) buffer systems are extracted and compared later in Fig. 8.3.

served in both solar cells. Since the surface and the cross section of the ZnO layer have different crystal orientations, the different CPD values are most likely due to the different surface dipole characteristics at these two positions. With the values of ΔCPD in the range of 450 – 500 mV from the former chapters as a reference, the CPD value on the ZnO surface seems to be more reasonable for forming the ΔCPD value.

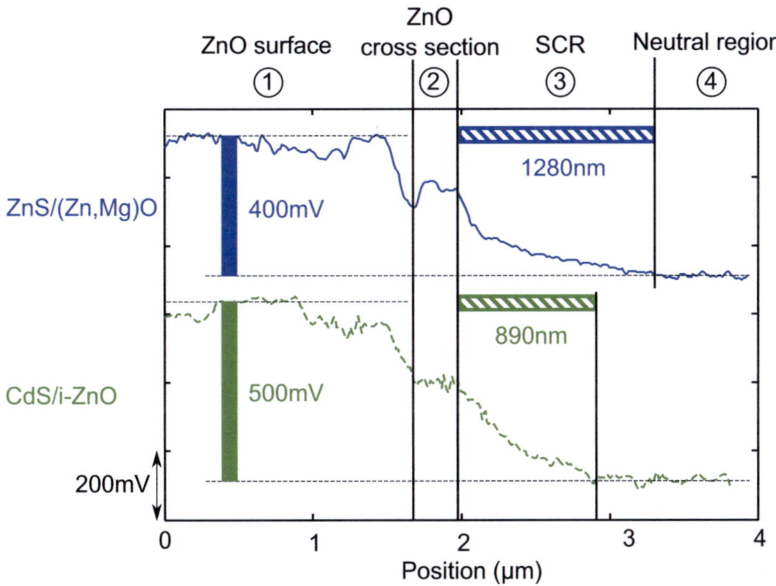

Figure 8.3: CPD line sections extracted from the marked positions in Fig. 8.2. They display the CPD distributions at heterojunctions of CIGS solar cells with CdS/i-ZnO (green dashed line) and ZnS/(Zn,Mg)O (blue solid line) buffer systems, respectively. Both curves can be separated into ① the surface of ZnO layer, ② the cross section of ZnO layer, ③ the space charge region (SCR) in CIGS and ④ the neutral region in CIGS. The comparison between the curves shows two major differences. First, ΔCPD in solar cell with the ZnS/(Zn,Mg)O buffer system (400 mV) is smaller than that in the solar cell with the CdS/i-ZnO buffer system (500 mV). Second, the SCR in solar cell with the ZnS/(Zn,Mg)O buffer system (1280 nm) is larger than that in the solar cell with the CdS/i-ZnO buffer system(890 nm).

With the band diagrams in Fig. 8.4 the physical meaning of the observations in KPFM measurements is explained. Since the ZnO layer is deposited at the final stage of the whole fabrication process, the electrical properties including the work function, the electron affinity and the ionization energy (valence band energy) of this layer is most likely the same in both samples. Therefore, the energy bands on the ZnO side are sketched identically. Based on this assumption, the energy bands on the CIGS side are derived. With the blue solid lines and the green dashed lines the situations in the ZnS/(Zn,Mg)O sample and the CdS/i-ZnO sample are indicated. Because ΔCPD is proportional to the work function difference between ZnO and CIGS and the work function of ZnO is considered as constant, the reduced ΔCPD value elucidates a reduced work function of CIGS. With a further assumption that the valence band energy E_V of CIGS is the same in both samples, it can be derived that the energy distance between the Fermi energy E_F and E_V is larger in the ZnS/(Zn,Mg)O sample. In general, the position of E_F over E_V is associated with the concentration of free charge carriers. In more detail, a larger energy distance is correlated to less free charge carriers. Consequently, a reduced value of ΔCPD suggests a smaller concentration of free charge carriers in the CIGS absorber of the ZnS/(Zn,Mg)O sample, which is purely a bulk property of CIGS. Also, the enlarged width of the SCR in the ZnS/(Zn,Mg)O sample can be explained by the smaller concentration of free charge carriers in CIGS. Besides, the band offset caused by interface dipoles can also contribute to the band alignment in a similar way. Thus, the enlarged SCR can be assigned to both the bulk property of the CIGS absorber and the interface property between CIGS and the buffer layer.

In a study of Witte *et al.* [79] ZnS/(Zn,Mg)O and CdS/i-ZnO samples that are similarly fabricated as in this work were compared. Electron beam induced current measurements in the junction configuration (J-EBIC) and capacitance voltage measurements were performed to extract the widths of the SCR and the acceptor densities of the samples. The authors found a

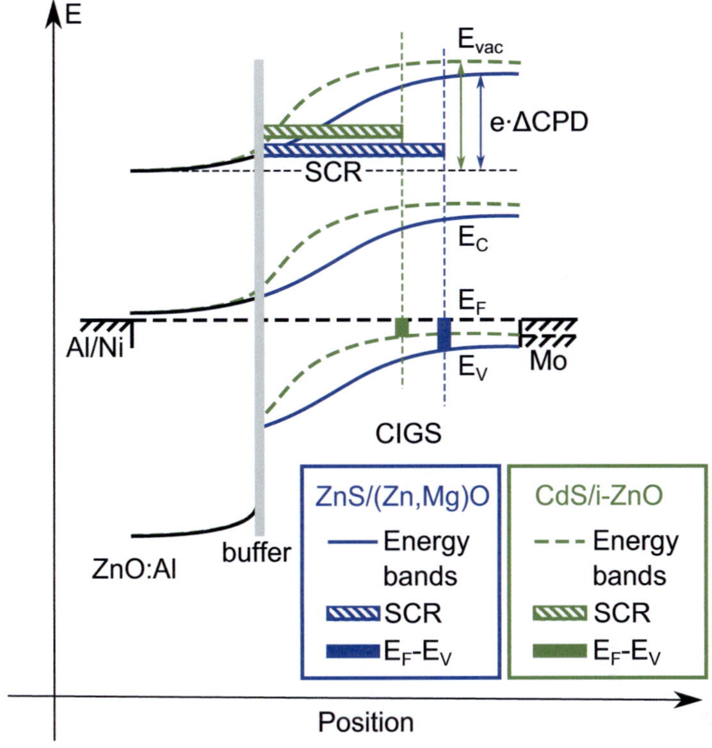

Figure 8.4: Energy band diagrams of CIGS solar cells with ZnS/(Zn,Mg)O (blue solid lines) and CdS/i-ZnO (green dashed lines) buffer systems can be derived on basis of the CPD data shown in Fig. 8.3. In the solar cell with the ZnS/(Zn,Mg)O buffer system, the smaller ΔCPD value indicates a larger energy gap between the Fermi energy E_F and the valence band energy E_V most likely due to a lower p-type doping density in the CIGS absorber. The larger SCR in this sample can be ascribed to both the lower doping density in CIGS and band offsets at the CIGS/buffer-interface.

larger SCR and a lower acceptor density in the ZnS/(Zn,Mg)O sample as the CdS/i-ZnO sample, which shows a high compatibility to the observations of the current work and can strengthen the estimations above.

Finally, one decisive question still has to be answered, i.e., how the deposition process of the buffer layer can change the bulk property of CIGS. The explanation may be the following: during the CBD process the CIGS layer is immersed into the chemical solution, which not only deposits a buffer layer onto the surface of CIGS but also changes the doping density in the bulk of CIGS most possibly by ion diffusion. In a very recent publication of Bastek et $al.$ [226] the diffusion behavior of Zn in $CuIn_{0.7}Ga_{0.3}Se_2$ layers fabricated at ZSW were reported. It was found that the diffusivity of Zn in $CuIn_{0.7}Ga_{0.3}Se_2$ layers is lower than that of Cd, due to the higher activation energy of Zn diffusion compared to that of Cd. Combing these findings with the observations of the current work, it is most probably that due to the lower diffusivity of Zn the CBD process for ZnS results in an underlying CIGS layer with a lower p-type doping density. As a direct consequence, the solar cell delivers a lower open circuit voltage.

8.3 Chapter conclusion

In this chapter a CIGS solar cell with the ZnS/(Zn,Mg)O buffer system was compared with a reference sample with the conventional CdS/i-ZnO buffer system. The CPD line section through heterojunction in the solar cell with the ZnS/(Zn,Mg)O buffer system showed a reduced value of the CPD difference between ZnO and CIGS, and a larger width of the space charge region in CIGS. These results indicated that the CBD process for ZnS and CdS not only microscopically modifies the interface conditions and macroscopically deposits a buffer layer but may also change the density of the p-type doping in the CIGS bulk material. Due to the lower diffusivity of Zn, the CBD process for ZnS most likely results in an underlying CIGS layer with a lower p-type doping density. This finding can well explain the generally observed reduction of the open circuit voltage in solar cells with the ZnS/(Zn,Mg)O buffer system.

9 Improving the solar cell performance with conclusions drawn in this work

In this chapter, the results in the last chapters are described and visualized with models. On that basis, a brief discussion about how the performance of CIGS-based and other kinds of thin-film solar cells could be further improved is evoked.

The investigation on CIGS solar cells with different Ga contents showed that the Ga content influences the distribution of acceptor and donor states in the CIGS absorber material. As explained with the energy band diagrams in Fig. 9.1 (a), the addition of Ga into $CuInSe_2$ leads to the generation of more acceptor states. Consequently, the Fermi energy in the CIGS material approaches the valence band energy. However, at a certain point (in this study [Ga]/([Ga]+[In])=0.63) the compensating donor states increases more rapidly. This leads to the reduction of the effective p-type doping of the absorber layer and the movement of the Fermi energy towards the middle of the band gap. As a result, the open circuit voltage of CIGS solar cells with a higher Ga content, particularly $CuGaSe_2$, is limited. It has been corroborated in numerous studies that Na can increase the free charge carrier concentration, presumably by eliminating the donor-type Cu on In antisites (In_{Cu}) [209, 227] or Se vacancies (V_{Se}) [210, 228]. Thus, as indicated in Fig. 9.1 (b), one way to enhance the p-type doping of the CIGS absorber material could be the deliberate introduction of other kinds of foreign atoms that hinder the formation of the compensating donor states. Furthermore, in agreement with some former work, this study showed a higher recombination rate in CIGS solar cells with increasing Ga content. Most recently, Contreras *et al.* published that the recombination process in CIGS solar

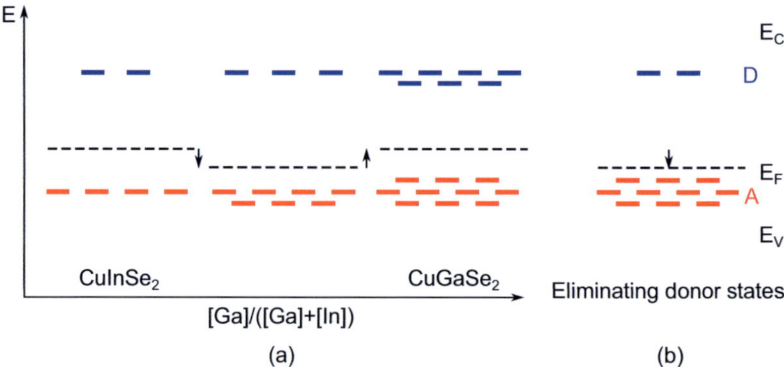

Figure 9.1: **(a)** By increasing the [Ga]/([Ga]+[In])-ratio more acceptor states (A) are generated in the CIGS absorber and consequently the Fermi energy approaches the valence band energy. However, at a certain point the compensating donor states (D) increases more rapidly. As a result, the effective p-type doping of the absorber is reduced and the Fermi energy moves towards the middle of the band gap. **(b)** The elimination the donor states will increase the p-type doping of the CIGS material and move the Fermi energy towards the valence band energy.

cells with higher Ga contents (band gap energy $1.2 - 1.45\,eV$) can be effectively reduced through the application of higher processing temperature up to $600 - 650\,°C$ [229]. The elevated temperature most likely improves the crystal quality of the CIGS absorber material and results in less deep defects that act as recombination centers. Therefore, another way to realize efficiency improvement is the development of advanced substrates, which are resistant to even higher processing temperature.

Grain boundaries were analyzed on the surface and on cross sections of a $CuIn_{0.7}Ga_{0.3}Se_2$ absorber film. Based on the results, three-dimensional models of the potential distribution of the grain structure can be built up as depicted in Fig. 9.2. Very importantly, the dip- and peak-shaped potential variations (GB_{dip} and GB_{peak}) were observed to reduce to the order of the thermal activation energy at room temperature. This outcome shows that beneficial effect of CIGS grain boundaries in terms of the collection and transport of free charge carriers is rather questionable. They are most likely not very recombination active, which already provides a good base for fabricating a reasonable photovoltaic device [12]. This interesting feature of CIGS materials can give a clear guidance in the optimization of thin-film solar cells based on other kinds of polycrystalline materials, particularly, kesterite materials. Special attention should be paid to the electrical properties of grain boundaries of these materials. If deep defect levels exist at grain boundaries and they are not reduced by the atomic relaxation in the grain boundary region, as estimated by Yan *et al.* for CIGS materials [74], additional measures, e.g., chemical treatments, will have to be undertaken, in order to passivate grain boundaries. In addition, even under the illumination comparable to the solar cell test condition, potential steps (GB_{step}) between certain grains in the order of 50 mV were still observable.

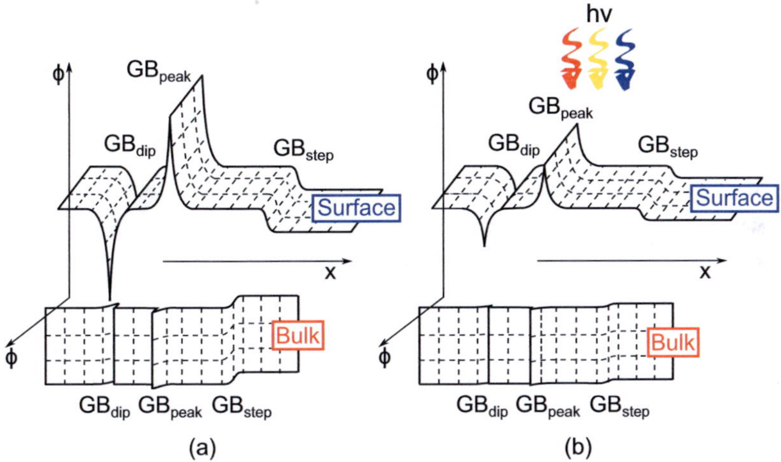

Figure 9.2: Three-dimensional models of potential distributions at grain boundaries in a CIGS absorber layer. **(a)** In darkness the magnitude of potential variations on the surface is much larger than that in the bulk. **(b)** Under illumination the potential variations at both positions are significantly reduced.

This observation could originate from surface effects like unsaturated surface dipoles or bulk effects such as compositional inhomogeneity between grains. In the latter case, there will be a potential fluctuation within the CIGS material. If the solar cell is considered to consist of a large number of mini solar cells that are formed by single grains, the total open circuit voltage will be limited by the lower values delivered by the mini solar cells. Hence, in the fabrication of CIGS solar cells with efficiencies surpassing the current record, the compositional inhomogeneity in the scale of grains should be carefully examined and avoided.

Due to a lower diffusivity of Zn in comparison to Cd, the underlying CIGS absorber material showed a lower p-type doping density. Therefore, a loss of the open circuit voltage in CIGS solar cells with the ZnS buffer layer was usually observed. As previously reported by different research groups [230, 231], the open circuit voltage can be improved by dipping of the CIGS absorber layer shortly into a water solution of NH_4OH and $CdSO_4$. This dip process is often referred to as the partial electrolyte (PE) treatment and results in the Cd diffusion into the absorber. The advantage of this process is that no CdS layer is formed. Therefore, for improving the open circuit voltage, the CIGS absorber layer can be dipped in PE before the CBD process of ZnS. Following this procedure the deficiency of the Zn doping could be compensated by Cd doping. Also, the performance of CIGS solar cells with other alternative buffer layers, e.g., ZnSe, $In(OH)_3$ and In_2S_3 could be noticeably raised in this way.

10 Summary and outlook

Thin-film solar cells based on CIGS absorber materials show the highest power conversion efficiency among all kinds of thin-film solar cells. The distribution of the electrostatic potential in and between the materials in the solar cell has obviously a major impact on the superior performance of the device. This thesis reported on imaging of the electrostatic potential on untreated cross sections of operating CIGS solar cells using Kelvin probe force microscopy. Four topics were comprehensively studied and the major results are elucidated in the following:

1. The potential distribution through the CIGS/CdS/ZnO-heterojunction in CIGS solar cells with [Ga]/([Ga]+[In])-ratios (GGI) from 0 to 1 were analyzed. The potential drop between CIGS and ZnO was systematically investigated, providing direct evidence for a Fermi energy shifting in CIGS absorber layers. The Fermi energy was observed to approach the valence band energy of CIGS for GGI-ratios between 0 and 0.63. This effect can be attributed to the Ga addition and the accompanying increase of the Na content. The increase of both elements enhances the effective p-type doping of the CIGS absorber. The Fermi energy was found to withdraw from the valence band energy for GGI-ratios over 0.63. This effect can be explained by the self-compensation effect, which leads to the Fermi energy shifting towards the middle of the band gap. Also, the slightly higher Cu content of the $CuGaSe_2$ sample may additionally contribute to this effect. Combining this outcome with results of the current density-voltage and external quantum efficiency measurements, the diffusion voltages of individual solar cells were deduced and compared with

the directly measured open circuit voltages. An increasing split-off between these two quantities was observed for Ga addition indicating a higher recombination rate of free charge carriers. This finding explained the unsatisfying performance of CIGS solar cells with higher Ga contents.

2. Potential variations at grain boundaries were analyzed on the surface and on untreated cross sections of the $CuIn_{0.7}Ga_{0.3}Se_2$ absorber in a high-efficiency solar cell. Differently to the previous studies, where only one single type or two types of potential variations at grain boundaries had been observed, three different types (dip-, step-, or peak-shape) were shown on samples in this work. Interestingly, the potential variations on cross sections were found to be much smaller than those on the surface of the same absorber. Since the properties of grain boundaries on cross sections can be expected to resemble more closely the ones buried in the bulk as those on absorber surfaces, the importance to interpret functionalities of grain boundaries based on observations on cross sections was pointed out. Moreover, the grain boundaries on cross sections were further observed under illumination comparable to the standard test conditions for solar cells. The potential variations at grain boundaries were found to decrease to less than 50 mV, which is much smaller than the potential drop induced by the solar cell heterojunction. All these findings suggested that CIGS grain boundaries do not play an active role in terms of charge carrier collection and transport.

3. The potential distribution through the heterojunction in a highly efficient solar cell based on $CuIn_{0.7}Ga_{0.3}Se_2$ absorber was analyzed under defined white light illumination. At the same time, the photovoltage and the photocurrent between the electrodes were recorded. It was found that the potential drop through the heterojunction decreases with increasing illumination intensity and vanishes over in-

tensities comparable to the standard test conditions for solar cells. Furthermore, the magnitude of the decrease of the potential drop correlated with the increase of the photovoltage and the photocurrent increased linearly with the illumination intensity. All these findings agreed well with the fundamental theory for solar cells and indicated that the cleaved solar cells are still fully functional and the cross sections are not severely dominated by surface states and dipoles. This outcome provided an important proof for the validity of the conclusions made in this work.

4. Potential distributions through the heterojunction in CIGS solar cells with the ZnS/(Zn,Mg)O and the conventional CdS/i-ZnO buffer systems were compared. The results showed a reduced value of the potential drop between ZnO and CIGS and a larger width of the space charge region in the solar cell with the ZnS/(Zn,Mg)O buffer system. It was concluded that due to the lower diffusivity of Zn the CBD process for ZnS results in a lower p-type doping density in the bulk of the underlying CIGS absorber layer. The generally observed reduction of the open circuit voltage of ZnS/(Zn,Mg)O samples could be well explained by this finding.

With these acquired knowledge, some recommendations on the optimization of thin-film solar cells based on CIGS and other kinds of polycrystalline materials were made at the end of this work, which would improve their efficiencies beyond the current records.

In the future work, KPFM measurements on cross sections of CIGS solar cells can be further improved. Since the cleavage of the samples and the KPFM measurements were carried out in air, the influence of the ambient conditions on the surface conditions like oxidation or water films between the tip and sample can be never totally excluded. In the next step, it is obligate to place the cleavage process and the KPFM setup into a glove box filled with inert gas. In this way the aforementioned surface effects can be effectively minimized, which will presumably lead to a higher potential contrast. Additionally, it was shown that the potential drop through the heterojunction is reduced by about $200\,\mathrm{mV}$ due to the detection laser. This problem can be resolved by exchanging the current cantilevers with the ones with broader beams or implementing a laser with a smaller spot. In this way, the laser light can be totally blocked. Another option is to apply cantilevers with longer beams. The laser spot can be positioned far away from the scanning point, so that the charge carriers generated by the laser spot cannot diffuse to the scanning point and have any influence on the potential drop.

Bibliography

[1] *Statistical review of world energy full report 2012*, BP, 2012, http://www.bp.com/statisticalreview.

[2] A. Bosio and A. Romero, Eds., *Thin film solar cells: current status and future trends.* New York: Nova Biomedical Books, 2010.

[3] *Entwicklung der erneuerbaren Energien in Deutschland im Jahr 2011*, Bundesministerium für Umwelt, Naturschutz und Reaktorsicherheit, 2012, http://www.bmu.de/service/publikationen/downloads/.

[4] *2010 solar technologies market report*, National Renewable Energy Laboratory, 2010, http://www.nrel.gov/docs/fy12osti/51847.pdf.

[5] L. Kazmerski, *NREL compilation of the best research solar cell efficiencies*, National Renewable Energy Laboratory, 2012, http://www.nrel.gov/ncpv/images/efficiency_chart.jpg.

[6] P. Jackson, D. Hariskos, E. Lotter, S. Paetel, R. Wuerz, R. Menner, W. Wischmann, and M. Powalla, "New world record efficiency for $Cu(In,Ga)Se_2$ thin-film solar cells beyond 20%," *Prog. Photovoltaics Res. Appl.*, vol. 19, pp. 894–897, 2011.

[7] O. Schultz, S. W. Glunz, and G. P. Willeke, "Multicrystalline silicon solar cells exceeding 20% efficiency," *Prog. Photovoltaics Res. Appl.*, vol. 12, pp. 553–558, 2004.

[8] M. A. Green, K. Emery, Y. Hishikawa, W. Warta, and E. D. Dunlop, "Solar cell efficiency tables (version 39)," *Prog. Photovoltaics Res. Appl.*, vol. 20, pp. 12–20, 2012.

[9] S. Niki, M. Contreras, I. Repins, M. Powalla, K. Kushiya, S. Ishizuka, and K. Matsubara, "CIGS absorbers and processes," *Prog. Photovoltaics Res. Appl.*, vol. 18, pp. 453–466, 2010.

[10] I. Repins, M. A. Contreras, B. Egaas, C. DeHart, J. Scharf, C. L. Perkins, B. To, and R. Noufi, "19.9%-efficient ZnO/CdS/CuInGaSe$_2$ solar cells with 81.2% fill factor," *Prog. Photovoltaics Res. Appl.*, vol. 16, pp. 235–239, 2008.

[11] L. Leamy, G. Pike, and C. Seager, Eds., *Grain boundaries in semiconductors*. New York: North-Holland, 1982.

[12] U. Rau, K. Taretto, and S. Siebentritt, "Grain boundaries in Cu(In,Ga)(Se,S)$_2$ thin-film solar cells," *Appl. Phys. A*, vol. 96, pp. 221–234, 2009.

[13] C. Jiang, R. Noufi, K. Ramanathan, J. A. AbuShama, H. R. Moutinho, and M. M. Al-Jassim, "Does the local built-in potential on grain boundaries of Cu(In,Ga)Se$_2$ thin films benefit photovoltaic performance of the device?" *Appl. Phys. Lett.*, vol. 85, pp. 2625–2627, 2004.

[14] J. B. Li, V. Chawla, and B. M. Clemens, "Investigating the role of grain boundaries in CZTS and CZTSSe thin film solar cells with scanning probe microscopy," *Adv. Mater.*, vol. 24, pp. 720–723, 2012.

[15] C. Jiang, F. S. Hasoon, H. R. Moutinho, H. A. Al-Thani, M. J. Romero, and M. M. Al-Jassim, "Direct evidence of a buried homojunction in Cu(In,Ga)Se$_2$ solar cells," *Appl. Phys. Lett.*, vol. 82, pp. 127–129, 2003.

[16] *Press release*, Solar Frontier, 2012, http://www.solar-frontier.com/eng/news/2012/.

[17] *Press release*, Manz AG, 2012, http://www.manz.com/media/news/archive/2012.

[18] J. L. Shay, S. Wagner, and H. M. Kasper, "Efficient $CuInSe_2$/CdS solar cells," *Appl. Phys. Lett.*, vol. 27, pp. 89–90, 1975.

[19] L. L. Kazmerski, F. R. White, and G. K. Morgan, "Thin film $CuInSe_2$/CdS heterojunction solar cells," *Appl. Phys. Lett.*, vol. 29, pp. 268–270, 1976.

[20] R. A. Mickelsen and W. S. Chen, *US patent no. 4335266*, 1982, http://patents.justia.com/1982/04335266.html.

[21] V. K. Kapur and U. V. Choudary, *US patent no. 4581108*, 1986, http://patents.justia.com/1986/04581108.html.

[22] H. W. Schock, "Solar cells based on $CuInSe_2$ and related compounds: recent progress in Europe," *Sol. Energy Mater. Sol. Cells*, vol. 34, pp. 19–26, 1994.

[23] D. S. Albin, J. J. Carapella, M. A. Contreras, A. M. Gabor, R. Noufi, and A. L. Tennant, *US patent no. 5436204*, 1995, http://www.wikipatents.com/US-Patent-5436204/.

[24] M. Powalla, "Highly efficient CIGS solar cells and modules with different buffers and substrates," in *presentations of the International Workshop on CIGS Solar Cell Technology, Berlin*, 2010.

[25] D. Hariskos, B. Fuchs, R. Menner, N. Naghavi, C. Hubert, D. Lincot, and M. Powalla, "The Zn(S,O,OH)/ZnMgO buffer in thin-film $Cu(In,Ga)(Se,S)_2$-based solar cells part II: magnetron sputtering of the ZnMgO buffer layer for in-line co-evaporated $Cu(In,Ga)Se_2$ solar cells," *Prog. Photovolt: Res. Appl.*, vol. 17, pp. 479–488, 2009.

[26] *ZSW annual report 2010*, Zentrum für Sonnenenergie- und Wasserstoff-Forschung Baden-Württemberg, 2010.

[27] G. Voorwinden, R. Kniese, and M. Powalla, "In-line Cu(In,Ga)Se$_2$ co-evaporation processes with graded band gaps on large substrates," *Thin Solid Films*, vol. 431-432, pp. 538–542, 2003.

[28] M. Powalla, G. Voorwinden, D. Hariskos, P. Jackson, and R. Kniese, "Highly efficient CIS solar cells and modules made by the co-evaporation process," *Thin Solid Films*, vol. 517, pp. 2111–2114, 2009.

[29] W. Witte, "Mikroskopische Inhomogenitäten und opto-elektrische Eigenschaften von Cu(In,Ga)Se$_2$-Schichten," Ph.D. dissertation, Universität Erlangen-Nürnberg, 2010.

[30] M. A. Contreras, T. Nakada, M. Hongo, A. O. Pudov, and J. R. Sites, "ZnO/ZnS(O,OH)/Cu(In,Ga)Se$_2$/Mo solar cells with 18.6% efficiency," in *proceedings of the 3rd World Conference on Photovoltaic Energy Conversion, Osaka*, 2003, pp. 570–573.

[31] H. J. Hovel, Ed., *Semiconductors and semimetals Volume 11 Solar cells*. New York, San Francisco and London: Academic Press, 1975.

[32] P. Würfel, *Physics of solar cells: from basic principles to advanced concepts*. Weinheim: Wiley-VCH, 2005.

[33] T. Markvart and L. Castaner, Eds., *Practical handbook of photovoltaics: fundamentals and applications*. Oxford: Elsevier, 2003.

[34] B. Dimmler, H. Dittrich, R. Menner, and H. W. Schock, "Performance and optimization of heterojunctions based on Cu(Ga,In)Se$_2$," in *proceedings of the 19th IEEE Photovoltaic Specialists Conference, New Orleans*, 1987, pp. 1454–1460.

[35] C. Jensen, D. Tarrant, J. Ermer, and G. Pollock, "The role of gallium in CuInSe$_2$ solar cells fabricated by a two stage method," in *proceedings of the 23th IEEE Photovoltaic Specialists Conference, Louisville*, 1993, p. 577.

[36] W. N. Shafarman, R. Klenk, and B. E. McCandless, "Device and material characterization of Cu(InGa)Se$_2$ solar cells with increasing band gap," *J. Appl. Phys.*, vol. 79, pp. 7324–7328, 1996.

[37] J. M. Stewart, W. S. Chen, W. E. Devaney, and R. A. Mickelsen, "Thin film polycrystalline CuIn$_{1-x}$Ga$_x$Se$_2$ solar cells," in *proceedings of the 7th International Conference on Ternary and Multinary Compounds, Pittsburgh*, 1986, p. 59.

[38] H. W. Schock, "Strategies for the development of multinary chalcopyrite based thin film solar cells," in *proceedings of the 12th European Photovoltaic Solar Energy Conference, Amsterdam*, 1994, p. 944.

[39] R. Herberholz, V. Nadenau, U. Rühle, C. Köble, H. W. Schock, and B. Dimmler, "Prospects of wide-gap chalcopyrites for thin film photovoltaic modules," *Solar Energy Materials and Solar Cells*, vol. 49, pp. 227–237, 1997.

[40] D. J. Schröder, J. L. Hernandez, G. D. Berry, and A. A. Rockett, "Hole transport and doping states in epitaxial CuIn$_{1-x}$Ga$_x$Se$_2$," *J. Appl. Phys.*, vol. 83, p. 1519, 1998.

[41] S. Wei, S. B. Zhang, and A. Zunger, "Effects of Ga addition to CuInSe$_2$ on its electronic, structural, and defect properties," *Appl. Phys. Lett.*, vol. 72, pp. 3199–3201, 1998.

[42] U. Rau, A. Jasenek, H. Schock, F. Engelhardt, and T. Meyer, "Electronic loss mechanisms in chalcopyrite based heterojunction solar cells," *Thin Solid Films*, vol. 361–362, pp. 298–302, 2000.

[43] A. Rockett, "The electronic effects of point defects in $Cu(In_xGa_{1-x})Se_2$," *Thin Solid Films*, vol. 361-362, pp. 330–337, 2000.

[44] A. Bauknecht, S. Siebentritt, J. Albert, and M. C. Lux-Steiner, "Radiative recombination via intrinsic defects in $Cu_xGa_ySe_2$," *J. Appl. Phys.*, vol. 89, pp. 4391–4400, 2001.

[45] G. Hanna, A. Jasenek, U. Rau, and H. W. Schock, "Influence of the Ga-content on the bulk defect densities of $Cu(In,Ga)Se_2$," *Thin Solid Films*, vol. 387, pp. 71–73, 2001.

[46] R. Klenk, "Characterisation and modelling of chalcopyrite solar cells," *Thin Solid Films*, vol. 387, pp. 135–140, 2001.

[47] C. Sommerhalter, S. Sadewasser, T. Glatzel, T. Matthes, A. Jäger-Waldau, and M. C. Lux-Steiner, "Kelvin probe force microscopy for the characterization of semiconductor surfaces in chalcopyrite solar cells," *Surf. Sci.*, vol. 482–485, pp. 1362–1367, 2001.

[48] S. Siebentritt, "Wide gap chalcopyrites: material properties and solar cells," *Thin Solid Films*, vol. 403-404, pp. 1–8, 2002.

[49] M. Turcu, O. Pakma, and U. Rau, "Interdependence of absorber composition and recombination mechanism in $Cu(In,Ga)(Se,S)_2$ heterojunction solar cells," *Appl. Phys. Lett.*, vol. 80, pp. 2598–2600, 2002.

[50] J. T. Heath, J. D. Cohen, W. N. Shafarman, D. X. Liao, and A. A. Rockett, "Effect of Ga content on defect states in $CuIn_{1-x}Ga_xSe_2$ photovoltaic devices," *Appl. Phys. Lett.*, vol. 80, pp. 4540–4542, 2002.

[51] M. Saad and A. Kassis, "Analysis of illumination-intensity-dependent j-V characteristics of $ZnO/CdS/CuGaSe_2$ single crystal

solar cells," *Solar Energy Materials and Solar Cells*, vol. 77, pp. 415–422, 2003.

[52] S. Siebentritt and S. Schuler, "Defects and transport in the wide gap chalcopyrite $CuGaSe_2$," *J. Phys. Chem. Solids*, vol. 64, pp. 1621–1626, 2003.

[53] M. Glöckler and J. R. Sites, "Efficiency limitations for wide-band-gap chalcopyrite solar cells," *Thin Solid Films*, vol. 480–481, pp. 241–245, 2005.

[54] M. Rusu, S. Doka, C. Kaufmann, N. Grigorieva, T. Schedel-Niedrig, and M. C. Lux-Steiner, "Solar cells based on CCSVT-grown $CuGaSe_2$-absorber and device properties," *Thin Solid Films*, vol. 480–481, pp. 341–346, 2005.

[55] S. Wei and S. B. Zhang, "Defect properties of $CuInSe_2$ and $CuGaSe_2$," *J. Phys. Chem. Solids*, vol. 66, pp. 1994–1999, 2005.

[56] M. Bär, M. Rusu, S. Lehmann, T. Schedel-Niedrig, I. Lauermann, and M. C. Lux-Steiner, "The chemical and electronic surface and interface structure of $CuGaSe_2$ thin-film solar cell absorbers," *Appl. Phys. Lett.*, vol. 93, p. 232104, 2008.

[57] S. Jung, S. Ahn, J. H. Yun, J. Gwak, D. Kim, and K. Yoon, "Effects of Ga contents on properties of CIGS thin films and solar cells fabricated by co-evaporation technique," *Curr. Appl Phys.*, vol. 10, pp. 990–996, 2010.

[58] M. Rusu, M. Bär, D. F. Marrón, S. Lehmann, T. Schedel-Niedrig, and M. Lux-Steiner, "Transport properties of $CuGaSe_2$-based thin-film solar cells as a function of absorber composition," *Thin Solid Films*, vol. 519, pp. 7304–7307, 2011.

[59] J. Serhan, Z. Djebbour, D. Mencaraglia, F. Couzini-Devy, N. Barreau, and J. Kessler, "Influence of Ga content on defects in $CuIn_xGa_{1-x}Se_2$ based solar cell absorbers investigated by sub-gap modulated photo-current and admittance spectroscopy," *Thin Solid Films*, vol. 519, pp. 7312–7316, 2011.

[60] S. Lany, Y. Zhao, C. Perrson, and A. Zunger, "Halogen n-type doping of chalcopyrite semiconductors," *Appl. Phys. Lett.*, vol. 86, p. 042109, 2005.

[61] S. Lany and A. Zunger, "Light- and bias-induced metastabilities in $Cu(In,Ga)Se_2$ based solar cells caused by the $(V_{Se}\text{-}V_{Cu})$ vacancy complex," *J. Appl. Phys.*, vol. 100, p. 113725, 2006.

[62] S. Siebentritt, M. Igalson, C. Persson, and S. Lany, "The electronic structure of chalcopyrites-bands, point defects and grain boundaries," *Prog. Photovoltaics Res. Appl.*, vol. 9999, pp. 390–410, 2010.

[63] R. Scheer and H.-W. Schock, Eds., *Chalcogenide photovoltaics*. Weinheim: WILEY-VCH, 2011.

[64] S. Schuler, S. Siebentritt, S. Nishiwaki, N. Rega, J. Beckmann, S. Brehme, and M. C. Lux-Steiner, "Self-compensation of intrinsic defects in the ternary semiconductor $CuGaSe_2$," *Phys. Rev. B*, vol. 69, p. 045210, 2004.

[65] A. Bauknecht, S. Siebentritt, A. Gerhard, W. Harneit, S. Brehme, J. Albert, S. Rushworth, and M. C. Lux-Steiner, "Defects in $CuGaSe_2$ thin films grown by MOCVD," *Thin Solid Films*, vol. 361–362, pp. 426–431, 2000.

[66] S. Siebentritt, A. Gerhard, S. Brehme, and M. C. Lux-Steiner, "Composition dependent doping and transport properties of $CuGaSe_2$," in *proceedings of MRS Spring Meeting, San Francisco*, vol. 668: H.4.4., 2001, pp. 1–6.

[67] W. Bollmann, Ed., *Crystal lattices, interfaces, matrices: an extension of crystallography.* Geneva: Polycrystal Book Service, 1982.

[68] C. Lei, C. M. Li, A. Rockett, and I. M. Robertson, "Grain boundary compositions in Cu(InGa)Se$_2$," *J. Appl. Phys.*, vol. 101, p. 024909, 2007.

[69] M. J. Hetzer, Y. M. Strzhemechny, M. Gao, M. A. Contreras, A. Zunger, and L. J. Brillson, "Direct observation of copper depletion and potential changes at copper indium gallium diselenide grain boundaries," *Appl. Phys. Lett.*, vol. 86, p. 162105, 2005.

[70] M. J. Romero, K. Ramanathan, M. A. Contreras, M. M. Al-Jassim, R. Noufi, and P. Sheldon, "Cathodoluminescence of Cu(In,Ga)Se$_2$ thin films used in high-efficiency solar cells," *Appl. Phys. Lett.*, vol. 83, pp. 4770–4772, 2003.

[71] D. Abou-Ras, C. Koch, V. Küstner, P. A. van Aken, U. Jahn, M. A. Contreras, R. Caballero, C. A. Kaufmann, R. Scheer, T. Unold, and H.-W. Schock, "Grain-boundary types in chalcopyrite-type thin films and their correlations with film texture and electrical properties," *Thin Solid Films*, vol. 517, pp. 2545–2549, 2009.

[72] O. Cojocaru-Miredin, P. Choi, R. Wuerz, and D. Raabe, "Atomic-scale characterization of the CdS/CuInSe$_2$ interface in thin-film solar cells," *Appl. Phys. Lett.*, vol. 98, p. 103504, 2011.

[73] S. Siebentritt, S. Sadewasser, M. Wimmer, C. Leendertz, T. Eisenbarth, and M. C. Lux-Steiner, "Evidence for a neutral grain-boundary barrier in chalcopyrites," *Phys. Rev. Lett.*, vol. 97, p. 146601, 2006.

[74] Y. Yan, C. Jiang, R. Noufi, S. Wei, H. R. Moutinho, and M. M. Al-Jassim, "Electrically benign behavior of grain boundaries in polycrystalline CuInSe$_2$ films," *Phys. Rev. Lett.*, vol. 99, p. 235504, 2007.

[75] R. Baier, D. Abou-Ras, T. Rissom, M. C. Lux-Steiner, and S. Sadewasser, "Symmetry-dependence of electronic grain boundary properties in polycrystalline CuInSe$_2$ thin films," *Appl. Phys. Lett.*, vol. 99, p. 172102, 2011.

[76] Z.-H. Zhang, X.-C. Tang, O. Kiowski, M. Hetterich, U. Lemmer, M. Powalla, and H. Hölscher, "Reevaluation of the beneficial effect of Cu(In,Ga)Se$_2$ grain boundaries using Kelvin probe force microscopy," *Appl. Phys. Lett.*, vol. 100, p. 203903, 2012.

[77] X. X. Liu and J. R. Sites, "Solar cell collection efficiency and its variation with voltage," *J. Appl. Phys.*, vol. 75, pp. 577–581, 1994.

[78] Y. Hamakawa, Ed., *Thin-film solar cells: next generation photovoltaics and its applications.* Berlin and Heidelberg: Springer, 2004.

[79] W. Witte, D. Hariskos, and M. Powalla, "Comparison of charge distributions in CIGS thin-film solar cells with ZnS/(Zn,Mg)O and CdS/i-ZnO buffers," *Thin Solid Films*, vol. 519, pp. 7549 – 7552, 2011.

[80] G. Zoppi, I. Forbes, R. W. Miles, P. J. Dale, J. J. Scragg, and L. M. Peter, "Cu$_2$ZnSnSe$_4$ thin film solar cells produced by selenisation of magnetron sputtered precursors," *Prog. Photovolt: Res. Appl.*, vol. 17, pp. 315–319, 2009.

[81] G. Binnig, H. Rohrer, C. Gerber, and E. Weibel, "Surface studies by scanning tunneling microscopy," *Phys. Rev. Lett.*, vol. 49, pp. 57–61, 1982.

[82] G. Binnig, C. F. Quate, and C. Gerber, "Atomic force microscope," *Phys. Rev. Lett.*, vol. 56, pp. 930–933, 1986.

[83] Y. Martin, C. C. Williams, and H. K. Wickramasinghe, "Atomic force microscope-force mapping and profiling on a sub 100 Å scale," *J. Appl. Phys.*, vol. 61, pp. 4723–4729, 1987.

[84] Y. Martin and H. K. Wickramasinghe, "Magnetic imaging by 'force microscopy' with 1000 Å," *Appl. Phys. Lett.*, vol. 50, pp. 1455–1457, 1987.

[85] Y. Martin, D. W. Abraham, and H. K. Wickramasinghe, "High-resolution capacitance measurement and potentiometry by force microscopy," *Appl. Phys. Lett.*, vol. 52, pp. 1103–1105, 1988.

[86] J. R. Matey and J. Blanc, "Scanning capacitance microscopy," *J. Appl. Phys.*, vol. 57, pp. 1437–1444, 1985.

[87] M. Nonnenmacher, M. P. O'Boyle, and H. K. Wickramasinghe, "Kelvin probe force microscopy," *Appl. Phys. Lett.*, vol. 58, pp. 2921–2923, 1991.

[88] Lord Kelvin, "Contact electricity of metals," *Philosophical Magazine Series 5*, vol. 46, pp. 82–120, 1898.

[89] W. A. Zisman, "A new method of measuring contact potential differences in metals," *Rev. Sci. Instrum.*, vol. 3, pp. 367–370, 1932.

[90] R. Garcia, *Amplitude modulation atomic force microscopy.* Weinheim: WILEY-VCH, 2010.

[91] H. Hölscher, D. Ebeling, and U. Schwarz, "Friction at atomic-scale surface steps: Experiment and theory," *Phys. Rev. Lett.*, vol. 101, p. 246105, 2008.

[92] R. D. Piner, J. Zhu, F. Xu, S. Hong, and C. A. Mirkin, "'Dip-Pen' nanolithography," *Science*, vol. 283, pp. 661–663, 1999.

[93] S. Sadewasser and T. Glatzel, Eds., *Kelvin probe force microscopy: measuring and compensating electrostatic forces.* Heidelberg, Dordrecht, London and New York: Springer, 2011.

[94] E. Meyer, H. J. Hug, and R. Bennewitz, Eds., *Scanning probe microscopy: the lab on a tip.* Berlin, Heidelberg and New York: Springer, 2004.

[95] R. García and R. Pérez, "Dynamic atomic force microscopy methods," *Surf. Sci. Rep.*, vol. 47, pp. 197–301, 2002.

[96] T. R. Albrecht, P. Grütter, D. Horne, and D. Rugar, "Frequency modulation detection using high-Q cantilevers for enhanced force microscope sensitivity," *J. Appl. Phys.*, vol. 69, pp. 668–673, 1991.

[97] N. W. Ashcroft and N. D. Mermin, *Festkörperphysik.* Berlin and München: Oldenbourg, 1976.

[98] P. Y. Yu and M. Cardona, *Fundamentals of semiconductors: physics and materials properties.* Heidelberg, Dordrecht, London and New York: Springer, 2010.

[99] X.-C. Tang, "Aufbau Kelvin-Sonden-Rasterkraftmikroskopie und ihre Anwendung auf CIGS Solarzellen," Master's thesis, Karlsruhe Institute of Technology, 2010.

[100] T. Glatzel, S. Sadewasser, and M. C. Lux-Steiner, "Amplitude or frequency modulation-detection in Kelvin probe force microscopy," *Appl. Surf. Sci.*, vol. 210, pp. 84–89, 2003.

[101] M. Meade, "Advances in lock-in amplifiers," *J. Phys. E: Sci. Instrum.*, vol. 15, pp. 395–403, 1982.

[102] S. M. Sze and Kwok K. Ng, Eds., *Physics of semiconductor devices.* Hoboken and New Jersey: John Wiley & Sons, Inc., 2007.

[103] L. Kronik and Y. Shapira, "Surface photovoltage phenomena: theory, experiment, and applications," *Surf. Sci. Rep.*, vol. 37, pp. 1–206, 1999.

[104] A. Chirila, S. Buecheler, F. Pianezzi, P. Bloesch, C. Gretener, A. R. Uhl, C. Fella, L. Kranz, J. Perrenoud, S. Seyrling, R. Verma, S. Nishiwaki, Y. E. Romanyuk, G. Bilger, and A. N. Tiwari, "Highly efficient $Cu(In,Ga)Se_2$ solar cells grown on flexible polymer films," *Nat. Mater.*, vol. 10, pp. 857–861, 2011.

[105] W. H. Brattain and W. Shockley, "Density of surface states on silicon deduced from contact potential measurements," *Phys. Rev.*, vol. 72, pp. 345–345, 1947.

[106] R. Williams, "Surface photovoltage measurements on cadmium sulfide," *J. Phys. Chem. Solids*, vol. 23, pp. 1057 – 1066, 1962.

[107] J. Shappir and A. Many, "The effect of oxygen adsorption on the surface barrier height of CdS," *Surf. Sci.*, vol. 14, pp. 169 – 180, 1969.

[108] A. Waxman, "Surface photovoltage measurements in vapour-deposited CdS," *Solid-State Electron.*, vol. 9, pp. 303 – 310, 1966.

[109] K. Okumura, "Photovoltaic effects at the interface between amorphous selenium and organic polymers," *J. Appl. Phys.*, vol. 45, pp. 5317–5323, 1974.

[110] I. Flinn and D. Emmony, "Surface measurements on n-type gallium arsenide," *Phys. Lett.*, vol. 6, pp. 133–135, 1963.

[111] W. Mönch, Ed., *Semiconductor surfaces and interfaces.* Berlin, Heidelberg and New York: Springer, 1995.

[112] J. M. Palau, "Fermi level pinning on (110) GaAs surfaces studied by CPD and SPV topographies," *J. Vac. Sci. Technol.*, vol. 19, p. 192, 1981.

[113] A. Ismail, A. Ben Brahim, J. Palau, and L. Lassabatere, "Comparison between GaAs(110) and InP(110) surface properties induced by cleavage defects and by oxygen adsorption," *Surf. Sci.*, vol. 162, pp. 195–201, 1985.

[114] L. Lassabatère, "Study of the GaAs-Au and Si-SiO$_2$ interface formation by the Kelvin method," *J. Vac. Sci. Technol., B*, vol. 1, pp. 540–545, 1983.

[115] G. P. Kochanski and R. Bell, "STM measurements of photovoltage on Si(111) and Si(111):Ge," *Surf. Sci.*, vol. 273, pp. L435 – L440, 1992.

[116] D. G. Cahill, "Scanning tunneling microscopy of photoexcited carriers at the Si(001) surface," *J. Vac. Sci. Technol., B*, vol. 9, pp. 564–567, 1991.

[117] D. Gorelik, S. Aloni, J. Eitle, D. Meyler, and G. Haase, "The role of adsorbed alkali metal atoms in the enhancement of surface reactivity: a scanning tunneling microscopy study of low coverage K/Si(111)7x7 surfaces," *J. Chem. Phys.*, vol. 108, pp. 9877–9884, 1998.

[118] W. Jägermann, "Adsorption of Br$_2$ on n-MoSe$_2$: modelling photoelectrochemistry in the UHV," *Chem. Phys. Lett.*, vol. 126, pp. 301–305, 1986.

[119] W. Jägermann, "Halogen adsorption on n-MoSe$_2$ (0001) van der Waals forces: simulation of electrochemical junctions in UHV," *Berichte der Bunsengesellschaft/Physical Chemistry Chemical Physics*, vol. 92, pp. 537–544, 1988.

[120] W. Jägermann and C. Pettenkofer, "Stability of photoelectrodes controlled by electronic factors: layered chalcogenides of group IVb," *Berichte der Bunsengesellschaft/Physical Chemistry Chemical Physics*, vol. 92, pp. 1354–1358, 1988.

[121] W. Jägermann, C. Pettenkofer, and B. Parkinson, "Cu and Ag deposition on layered p-type WSe_2: approaching the schottky limit," *Phys. Rev. B*, vol. 42, pp. 7487–7496, 1990.

[122] M. Sander, W. Jägermann, and H. J. Lewerenz, "Site-specific surface interaction of adsorbed water and halogens on copper indium selenide ($CuInSe_2$) surfaces," *J. Phys. Chem.*, vol. 96, pp. 782–790, 1992.

[123] C. Heske, G. Richter, Z. Chen, R. Fink, E. Umbach, W. Riedl, and F. Karg, "Influence of Na and H_2O on the surface properties of $Cu(In,Ga)Se_2$ thin films," *J. Appl. Phys.*, vol. 82, pp. 2411–2420, 1997.

[124] A. Meeder, L. Weinhardt, R. Stresing, D. Fuertes Marrón, R. Würz, S. Babu, T. Schedel-Niedrig, M. Lux-Steiner, C. Heske, and E. Umbach, "Surface and bulk properties of $CuGaSe_2$ thin films," *J. Phys. Chem. Solids*, vol. 64, pp. 1553–1557, 2003.

[125] H. Neumann, M. V. Yakushev, and R. D. Tomlinson, "Diffusion effects at the Au/p-$CuInSe_2$ contact studied by XPS," *Cryst. Res. Technol.*, vol. 38, pp. 676–683, 2003.

[126] M. Bruening, E. Moons, D. Yaron-Marcovich, D. Cahen, J. Libman, and A. Shanzer, "Polar ligand adsorption controls semiconductor surface potentials," *J. Am. Chem. Soc.*, vol. 116, pp. 2972–2977, 1994.

[127] M. Bruening, E. Moons, D. Cahen, and A. Shanzer, "Controlling the work function of CdSe by chemisorption of benzoic acid deriva-

tives and chemical etching," *J. Phys. Chem.*, vol. 99, pp. 8368–8373, 1995.

[128] S. Bastide, R. Butruille, D. Cahen, A. Dutta, J. Libman, A. Shanzer, L. Sun, and A. Vilan, "Controlling the work function of GaAs by chemisorption of benzoic acid derivatives," *J. Phys. Chem. B*, vol. 101, pp. 2678–2684, 1997.

[129] M. Bruening, R. Cohen, J. F. Guillemoles, T. Moav, J. Libman, A. Shanzer, and D. Cahen, "Simultaneous control of surface potential and wetting of solids with chemisorbed multifunctional ligands," *J. Am. Chem. Soc.*, vol. 119, pp. 5720–5728, 1997.

[130] R. Cohen, S. Bastide, D. Cahen, J. Libman, A. Shanzer, and Y. Rosenwaks, "Controlling electronic properties of CdTe by adsorption of dicarboxylic acid derivatives: relating molecular parameters to band bending and electron affinity changes," *Adv. Mater.*, vol. 9, pp. 746–749, 1997.

[131] R. Cohen, N. Zenou, D. Cahen, and S. Yitzchaik, "Molecular electronic tuning of Si surfaces," *Chem. Phys. Lett.*, vol. 279, pp. 270–274, 1997.

[132] R. Cohen, S. Bastide, D. Cahen, J. Libman, A. Shanzer, and Y. Rosenwaks, "Controlling surfaces and interfaces of semiconductors using organic molecules," *Opt. Mater.*, vol. 9, pp. 394–400, 1998.

[133] B. Goldstein and D. J. Szostak, "Surface photovoltage, band-bending and surface states on a-Si:H," *Surf. Sci.*, vol. 99, pp. 235–258, 1980.

[134] J. Szuber, "Surface photovoltage spectroscopy investigations of the electronic surface states on clean and oxygen-exposed polar GaAs(100) and GaAs(111) surfaces," *J. Electron. Spectrosc. Relat. Phenom.*, vol. 53, pp. 19–28, 1990.

[135] H. Lüth, M. Büchel, R. Dorn, M. Liehr, and R. Matz, "Electronic structure of cleaved clean and oxygen-covered GaAs(110) surfaces," *Phys. Rev. B*, vol. 15, pp. 865–874, 1977.

[136] S. Thurgate, K. Blight, and T. Laceusta, "Surface photovoltage studies of n-type and p-type InP," *Surf. Sci.*, vol. 310, pp. 103–112, 1994.

[137] C. Heske, R. Fink, E. Umbach, W. Riedl, and F. Karg, "Na-induced effects on the electronic structure and composition of $Cu(In,Ga)Se_2$ thin-film surfaces," *Appl. Phys. Lett.*, vol. 68, pp. 3431–3433, 1996.

[138] U. Rau, D. Braunger, R. Herberholz, H. W. Schock, J. Guillemoles, L. Kronik, and D. Cahen, "Oxygenation and air-annealing effects on the electronic properties of $Cu(In,Ga)Se_2$ films and devices," *J. Appl. Phys.*, vol. 86, pp. 497–505, 1999.

[139] P. A. Thiel and T. E. Madey, "The interaction of water with solid surfaces: fundamental aspects," *Surf. Sci. Rep.*, vol. 7, pp. 211–385, 1987.

[140] M. A. Henderson, "The interaction of water with solid surfaces: fundamental aspects revisited," *Surf. Sci. Rep.*, vol. 46, pp. 1–308, 2002.

[141] S. Ono, M. Takeuchi, and T. Takahashi, "Kelvin probe force microscopy on InAs thin films grown on GaAs giant step structures formed on (110)GaAs vicinal substrates," *Appl. Phys. Lett.*, vol. 78, pp. 1086–1088, 2001.

[142] H. Sugimura, Y. Ishida, K. Hayashi, O. Takai, and N. Nakagiri, "Potential shielding by the surface water layer in Kelvin probe force microscopy," *Appl. Phys. Lett.*, vol. 80, pp. 1459–1461, 2002.

[143] C. Sommerhalter, "Kelvinsondenkraftmikroskopie im Ultrahochvakuum zur Charakterisierung von Halbleiter-Heterodioden auf der Basis von Chalkopyriten," Ph.D. dissertation, Freie Universität Berlin, 1999.

[144] S. Sadewasser, T. Glatzel, S. Schuler, S. Nishiwaki, R. Kaigawa, and M. C. Lux-Steiner, "Kelvin probe force microscopy for the nano scale characterization of chalcopyrite solar cell materials and devices," *Thin Solid Films*, vol. 431-432, pp. 257–261, 2003.

[145] C. Jiang, R. Noufi, J. A. AbuShama, K. Ramanathan, H. R. Moutinho, J. Pankow, and M. M. Al-Jassim, "Local built-in potential on grain boundary of Cu(In,Ga)Se$_2$ thin films," *Appl. Phys. Lett.*, vol. 84, pp. 3477–3479, 2004.

[146] D. F. Marrón, S. Sadewasser, A. Meeder, T. Glatzel, and M. C. Lux-Steiner, "Electrical activity at grain boundaries of Cu(In,Ga)Se$_2$ thin films," *Phys. Rev. B*, vol. 71, p. 033306, 2005.

[147] D. F. Marrón, A. Meeder, S. Sadewasser, R. Wuerz, C. A. Kaufmann, T. Glatzel, T. Schedel-Niedrig, and M. C. Lux-Steiner, "Lift-off process and rear-side characterization of CuGaSe$_2$ chalcopyrite thin films and solar cells," *J. Appl. Phys.*, vol. 97, p. 094915, 2005.

[148] S. Sadewasser, "Surface potential of chalcopyrite films measured by KPFM," *Phys. Status Solidi A*, vol. 203, pp. 2571–2580, 2006.

[149] S. Sadewasser, "Microscopic characterization of individual grain boundaries in Cu-III-VI$_2$ chalcopyrites," *Thin Solid Films*, vol. 515, pp. 6136–6141, 2007.

[150] G. Hanna, T. Glatzel, S. Sadewasser, N. Ott, H. Strunk, U. Rau, and J. Werner, "Texture and electronic activity of grain boundaries in Cu(In,Ga)Se$_2$ thin films," *Appl. Phys. A*, vol. 82, pp. 1–7, 2006.

[151] S. Sadewasser, D. Abou-Ras, D. Azulay, R. Baier, I. Balberg, D. Cahen, S. Cohen, K. Gartsman, K. Ganesan, J. Kavalakkatt, W. Li, O. Millo, T. Rissom, Y. Rosenwaks, H. Schock, A. Schwarzman,

and T. Unold, "Nanometer-scale electronic and microstructural properties of grain boundaries in Cu(In,Ga)Se$_2$," *Thin Solid Films*, vol. 519, pp. 7341–7346, 2011.

[152] Y. Yan, C. Jiang, R. Noufi, S. Wei, H. R. Moutinho, and M. M. Al-Jassim, "Electrically benign behavior of grain boundaries in polycrystalline CuInSe$_2$ films," *Phys. Rev. Lett.*, vol. 99, p. 235504, 2007.

[153] C. Jiang, R. Noufi, K. Ramanathan, H. R. Moutinho, and M. M. Al-Jassim, "Electrical modification in Cu(In,Ga)Se$_2$ thin films by chemical bath deposition process of CdS films," *J. Appl. Phys.*, vol. 97, p. 053701, 2005.

[154] M. Rusu, T. Glatzel, A. Neisser, C. A. Kaufmann, S. Sadewasser, and M. C. Lux-Steiner, "Formation of the physical vapor deposited CdS/Cu(In,Ga)Se$_2$ interface in highly efficient thin film solar cells," *Appl. Phys. Lett.*, vol. 88, p. 143510, 2006.

[155] T. Glatzel, M. Rusu, S. Sadewasser, and M. C. Lux-Steiner, "Surface photovoltage analysis of thin CdS layers on polycrystalline chalcopyrite absorber layers by Kelvin probe force microscopy," *Nanotechnol.*, vol. 19, p. 145705, 2008.

[156] F. Streicher, S. Sadewasser, and M. C. Lux-Steiner, "Surface photovoltage spectroscopy in a Kelvin probe force microscope under ultrahigh vacuum," *Rev. Sci. Instrum.*, vol. 80, p. 013907, 2009.

[157] F. Streicher, S. Sadewasser, T. Enzenhofer, H. Schock, and M. C. Lux-Steiner, "Locally resolved surface photo voltage spectroscopy on Zn-doped CuInS$_2$ polycrystalline thin films," *Thin Solid Films*, vol. 517, pp. 2349–2352, 2009.

[158] A. Kikukawa, S. Hosaka, and R. Imura, "Silicon pn junction imaging and characterizations using sensitivity enhanced Kelvin probe force microscopy," *Appl. Phys. Lett.*, vol. 66, pp. 3510–3512, 1995.

[159] T. Meoded, R. Shikler, N. Fried, and Y. Rosenwaks, "Direct measurement of minority carriers diffusion length using Kelvin probe force microscopy," *Appl. Phys. Lett.*, vol. 75, pp. 2435–2437, 1999.

[160] S. Saraf and Y. Rosenwaks, "Local measurement of semiconductor band bending and surface charge using Kelvin probe force microscopy," *Surf. Sci.*, vol. 574, pp. L35 – L39, 2005.

[161] M. Tanimoto and O. Vatel, "Kelvin probe force microscopy for characterization of semiconductor devices and processes," *J. Vac. Sci. Technol. B*, vol. 14, pp. 1547–1551, 1996.

[162] U. T, M. Arakawa, S. Kishimoto, T. Mizutani, T. Kagawa, and H. Iwamura, "Cross-sectional potential imaging of compound semiconductor heterostructure by Kelvin probe force microscopy," *Jpn. J. Appl. Phys.*, vol. 37, pp. 1522–1526, 1999.

[163] T. Mizutani, T. Usunami, S. Kishimoto, and K. Maezawa, "Measurement of contact potential of GaAs pn junctions by Kelvin probe force microscopy," *Jpn. J. Appl. Phys.*, vol. 38, pp. 4893–4894, 1999.

[164] T. Mizutani, T. Usunami, S. Kishimoto, and K. Maezawa, "Measurement of contact potential of GaAs/AlGaAs heterostructure using Kelvin probe force microscopy," *Jpn. J. Appl. Phys.*, vol. 38, pp. L767–L769, 1999.

[165] F. Robin, H. Jacobs, O. Homan, A. Stemmer, and W. Bächtold, "Investigation of the cleaved surface of a p-i-n laser using Kelvin probe force microscopy and two dimensional physical simulations," *Appl. Phys. Lett.*, vol. 76, pp. 2907–2909, 2000.

[166] A. Schwarzman, E. Grunbaum, E. Strassburg, E. Lepkifker, A. Boag, Y. Rosenwaks, T. Glatzel, Z. Barkay, M. Mazzer, and K. Barnham,

"Nanoscale potential distribution across multiquantum well structures: Kelvin probe force microscopy and secondary electron imaging," *J. Appl. Phys.*, vol. 98, p. 084310, 2005.

[167] R. Shikler, T. Meoded, N. Fried, and Y. Rosenwaks, "Potential imaging of operating light-emitting devices using Kelvin force microscopy," *Appl. Phys. Lett.*, vol. 74, pp. 2972–2974, 1999.

[168] R. Shikler, T. Meoded, N. Fried, B. Mishori, and Y. Rosenwaks, "Two-dimensional surface band structure of operating light emitting devices," *J. Appl. Phys.*, vol. 86, pp. 107–113, 1999.

[169] G. Lévêque, P. Girard, E. Skouri, and D. Yarekha, "Measurements of electric potential in a laser diode by Kelvin probe force microscopy," *Appl. Surf. Sci.*, vol. 157, pp. 251–255, 2000.

[170] A. V. Ankudinov, V. P. Evtikhiev, E. Y. Kotelnikov, A. N. Titkov, and R. Laiho, "Voltage distributions and nonoptical catastrophic mirror degradation in high power InGaAs/AlGaAs/GaAs lasers studied by Kelvin probe force microscopy," *J. Appl. Phys.*, vol. 93, pp. 432–437, 2003.

[171] T. Mizutani, M. Arakawa, and S. Kishimoto, "Potential profile measurement of cleaved surface of GaAs HEMTs by Kelvin probe force microscopy," in *proceedings in Electron Devices Meeting*, 1996, pp. 31–34.

[172] T. Mizutani, M. Arakawa, and S. Kishimoto, "Two-dimensional potential profile measurement of GaAs HEMT's by Kelvin probe force microscopy," *IEEE Electron Device Lett.*, vol. 18, pp. 423–425, 1997.

[173] M. Arakawa, S. Kishimoto, and T. Mizutani, "Kelvin probe force microscopy for potential distribution measurement of cleaved sur-

face of GaAs devices," *Jpn. J. Appl. Phys.*, vol. 36, pp. 1826–1829, 1997.

[174] K. Matsunami, T. Takeyama, T. Usunami, S. Kishimoto, K. Maezawa, T. Mizutani, M. Tomizawa, P. Schmid, K. M. Lipka, and E. Kohn, "Potential profile measurements of GaAs MESFETs passivated with low-temperature grown GaAs layer by Kelvin probe force microscopy," *Solid-State Electron.*, vol. 43, pp. 1547–1553, 1999.

[175] L. Bürgi, T. J. Richards, R. H. Friend, and H. Sirringhaus, "Close look at charge carrier injection in polymer field-effect transistors," *J. Appl. Phys.*, vol. 94, pp. 1547–1553, 2003.

[176] C.-S. Jiang, H. R. Moutinho, J. F. Geisz, D. J. Friedman, and M. M. Al-Jassim, "Direct measurement of electrical potentials in GaInP$_2$ solar cells," *Appl. Phys. Lett.*, vol. 81, pp. 2569–2571, 2002.

[177] C.-S. Jiang, D. Friedman, J. F. Geisz, H. R. Moutinho, M. J. Romero, and M. M. Al-Jassim, "Distribution of built-in electrical potential in GaInP$_2$/GaAs tandem-junction solar cells," *Appl. Phys. Lett.*, vol. 83, pp. 1572–1574, 2003.

[178] C.-S. Jiang, H. R. Moutinho, R. Reedy, M. M. Al-Jassim, and A. Blosse, "Two-dimensional junction identification in multicrystalline silicon solar cells by scanning Kelvin probe force microscopy," *J. Appl. Phys.*, vol. 104, p. 104501, 2008.

[179] H. R. Moutinho, R. G. Dhere, C. Jiang, Y. Yan, D. S. Albin, and M. M. Al-Jassim, "Investigation of potential and electric field profiles in cross sections of CdTe/CdS solar cells using scanning Kelvin probe microscopy," *J. Appl. Phys.*, vol. 108, p. 074503, 2010.

[180] T. Glatzel, D. F. Marrón, T. Schedel-Niedrig, S. Sadewasser, and M. C. Lux-Steiner, "CuGaSe$_2$ solar cell cross section studied by

Kelvin probe force microscopy in ultrahigh vacuum," *Appl. Phys. Lett.*, vol. 81, pp. 2017–2019, 2002.

[181] D. F. Marrón, T. Glatzel, A. Meeder, T. Schedel-Niedrig, S. Sadewasser, and M. C. Lux-Steiner, "Electronic structure of secondary phases in Cu-rich CuGaSe$_2$ solar cell devices," *Appl. Phys. Lett.*, vol. 85, pp. 3755–3757, 2004.

[182] T. Glatzel, H. Steigert, S. Sadewasser, R. Klenk, and M. Lux-Steiner, "Potential distribution of Cu(In,Ga)(S,Se)$_2$-solar cell cross-sections measured by Kelvin probe force microscopy," *Thin Solid Films*, vol. 480-481, pp. 177–182, 2005.

[183] R. Mainz, F. Streicher, D. Abou-Ras, S. Sadewasser, R. Klenk, and M. C. Lux-Steiner, "Combined analysis of spatially resolved electronic structure and composition on a cross-section of a thin film Cu(In$_{1-x}$Ga$_x$)S$_2$ solar cell," *Phys. Status Solidi A*, vol. 206, pp. 1017–1020, 2009.

[184] T. Glatzel, "Kelvinsondenkraftmikroskopie am Heteroübergang (Zn,Mg)O/Cu(In,Ga)(S,Se)$_2$-Chalkopyrit," Ph.D. dissertation, Freie Universität Berlin, 2003.

[185] D. Schmid, M. Ruckh, F. Grunwald, and H. W. Schock, "Chalcopyrite/defect chalcopyrite heterojunctions on the basis of CuInSe$_2$," *J. Appl. Phys.*, vol. 73, pp. 2902–2909, 1993.

[186] A. Klein and W. Jägermann, "Fermi-level-dependent defect formation in Cu-chalcopyrite semiconductors," *Appl. Phys. Lett.*, vol. 74, pp. 2283–2285, 1999.

[187] Y. Yan, K. M. Jones, J. Abushama, M. Young, S. Asher, M. M. Al-Jassim, and R. Noufi, "Microstructure of surface layers in Cu(In,Ga)Se$_2$ thin films," *Appl. Phys. Lett.*, vol. 81, pp. 1008–1010, 2002.

[188] S. H. Kwon, S. C. Park, B. T. Ahn, K. H. Yoon, and J. Song, "Effect of $CuIn_3Se_5$ layer thickness on $CuInSe_2$ thin films and devices," *Solar Energy*, vol. 64, pp. 55–60, 1998.

[189] S. B. Zhang, S. Wei, and A. Zunger, "Stabilization of ternary compounds via ordered arrays of defect pairs," *Phys. Rev. Lett.*, vol. 78, pp. 4059–4062, 1997.

[190] Y. Rosenwaks, R. Shikler, T. Glatzel, and S. Sadewasser, "Kelvin probe force microscopy of semiconductor surface defects," *Phys. Rev. B*, vol. 70, p. 085320, 2004.

[191] D. Ziegler, J. Rychen, N. Naujoks, and A. Stemmer, "Compensating electrostatic forces by single-scan Kelvin probe force microscopy," *Nanotechnol.*, vol. 18, p. 225505, 2007.

[192] *Scanning Probe Microscopy Training Notebook Verion 3.0*, Digital Instruments.

[193] *AFM probe type: PPP-FMR*, NANOSENSORS^{TM}, http://www.nanoandmore.com/AFM-Probe-PPP-FMR.html.

[194] *MultiMode iC Manual v2.4*, nPoint.

[195] *C300 DSP Controller Manual v1.4*, nPoint.

[196] D. Liao and A. Rockett, "Cu depletion at the $CuInSe_2$ surface," *Appl. Phys. Lett.*, vol. 82, pp. 2829–2831, 2003.

[197] S. Sadewasser, T. Glatzel, R. Shikler, Y. Rosenwaks, and M. C. Lux-Steiner, "Resolution of Kelvin probe force microscopy in ultrahigh vacuum: comparison of experiment and simulation," *Appl. Surf. Sci.*, vol. 210, pp. 32–36, 2003.

[198] U. Zerweck, C. Loppacher, T. Otto, S. Grafström, and L. M. Eng, "Accuracy and resolution limits of Kelvin probe force microscopy," *Phys. Rev. B*, vol. 71, p. 125424, 2005.

[199] J. Schmutz, M. M. Schärfer, and H. Hölscher, "Colloid probes with increased tip height for higher sensitivity in friction force microscopy and less cantilever damping in dynamic force microscopy," *Rev. Sci. Instrum.*, vol. 79, p. 026103, 2008.

[200] C. Sommerhalter, T. Glatzel, T. W. Matthes, A. Jäger-Waldau, and M. C. Lux-Steiner, "Kelvin probe force microscopy in ultra high vacuum using amplitude modulation detection of the electrostatic forces," *Appl. Surf. Sci.*, vol. 157, pp. 263–268, 2000.

[201] P. M. Vilarinho, Y. Rosenwaks, and A. I. Kingon, Eds., *Scanning probe microscopy: characterization, nanofabrication and device application of functional materials.* Dordrecht: Kluwer Academic Publishers, 2005.

[202] H. Diesinger, D. Deresmes, J. Nys, and T. Mélin, "Kelvin force microscopy at the second cantilever resonance: an out-of-vacuum crosstalk compensation setup," *Ultramicroscopy*, vol. 108, pp. 773–781, 2008.

[203] H. B. Michaelson, "The work function of the elements and its periodicity," *J. Appl. Phys.*, vol. 48, pp. 4729–4733, 1977.

[204] T. Dullweber, O. Lundberg, J. Malmström, M. Bodegard, L. Stolt, U. Rau, H. W. Schock, and J. H. Werner, "Back surface band gap gradings in Cu(In,Ga)Se$_2$ solar cells," *Thin Solid Films*, vol. 387, pp. 11–13, 2001.

[205] H. O. Jacobs, P. Leuchtmann, O. J. Homan, and A. Stemmer, "Resolution and contrast in Kelvin probe force microscopy," *J. Appl. Phys.*, vol. 84, pp. 1168–1173, 1998.

[206] G. Elias, T. Glatzel, E. Meyer, and A. Schwarzman, "The role of the cantilever in Kelvin probe force microscopy measurements," *Beilstein J. Nanotechnol.*, vol. 2, pp. 252–260, 2011.

[207] D. J. Schröder and A. A. Rockett, "Electronic effects of sodium in epitaxial $CuIn_{1-x}Ga_xSe_2$," *J. Appl. Phys.*, vol. 82, pp. 4982–4985, 1997.

[208] D. Cahen, E. Moons, L. Chernyak, I. Lyubomirski, M. Bruening, A. Shanzer, and J. Libman, in *proceedings of the 5th International Symposium for Uses of Se and Te, Brussels*, 1994, p. 207.

[209] S. Wei, S. B. Zhang, and A. Zunger, "Effects of Na on the electrical and structural properties of $CuInSe_2$," *J. Appl. Phys.*, vol. 85, pp. 7214–7218, 1999.

[210] L. Kronik, D. Cahen, and H. Schock, "Effects of sodium on polycrystalline $Cu(In,Ga)Se_2$ and its solar cell performance," *Adv. Mater.*, vol. 10, pp. 31–35, 1998.

[211] S. Ishizuka, A. Yamada, M. M. Islam, H. Shibata, P. Fons, T. Sakurai, K. Akimoto, and S. Niki, "Na-induced variations in the structural, optical, and electrical properties of $Cu(In,Ga)Se_2$ thin films," *J. Appl. Phys.*, vol. 106, p. 034908, 2009.

[212] S. B. Zhang, S.-H. Wei, and A. Zunger, "Defect physics of the $CuInSe_2$ chalcopyrite semiconductor," *Phys. Rev. B*, vol. 57, pp. 9642–9656, 1998.

[213] T. Markvart and L. Castaner, Eds., *Practical handbook of photovoltaics: fundamentals and applications.* Oxford: Elsevier, 2003.

[214] P. Würfel, Ed., *Physics of solar cells.* Weinheim: WILEY-VCH, 2005.

[215] A. Rockett, "Performance-limitations in $Cu(In,Ga)Se_2$-based heterojunction solar cells," in *proceedings of 29th IEEE Photovoltaic Specialists Conference, New Orleans*, 2002, pp. 587–591.

[216] L. J. Brillson and Y. Lu, "ZnO schottky barriers and Ohmic contacts," *J. Appl. Phys.*, vol. 109, p. 121301, 2011.

[217] A. Chirilă, S. Buecheler, F. Pianezzi, P. Bloesch, C. Gretener, A. R. Uhl, C. Fella, L. Kranz, J. Perrenoud, S. Seyrling, R. Verma, S. Nishiwaki, Y. E. Romanyuki, G. Bilger, and A. N. Tiwari, "Highly efficient Cu(In,Ga)Se$_2$ solar cells grown on flexible polymer films," *Nat. Mater.*, vol. 10, pp. 857–861, 2011.

[218] D. K. Rao, J. J. B. Prasad, D. Sridevi, K. V. Reddy, and J. Sobhanadri, "Properties of p-CuInSe$_2$/Al schottky devices," *Phys. Status Solidi A*, vol. 94, pp. K153–K158, 1986.

[219] N. Jensen, U. Rau, R. M. Hausner, S. Uppal, and L. Oberbeck, "Recombination mechanisms in amorphous silicon/crystalline silicon heterojunction solar cells," *J. Appl. Phys.*, vol. 87, pp. 2639–2645, 2000.

[220] S. Sadewasser, T. Glatzel, M. Rusu, A. Jäger-Waldau, and M. C. Lux-Steiner, "High-resolution work function imaging of single grains of semiconductor surfaces," *Appl. Phys. Lett.*, vol. 80, pp. 2979–2981, 2002.

[221] W. Shockley and H. Queisser, "Detailed balance limit of efficiency of p-n junction solar cells," *J. Appl. Phys.*, vol. 97, pp. 510–519, 1961.

[222] K. Maturová, M. Kemerink, M. Wienk, D. Charrier, and R. Janssen, "Scanning Kelvin probe microscopy on bulk heterojunction polymer blends," *Adv. Funct. Mater.*, vol. 19, pp. 1379–1386, 2009.

[223] C. Hubert, N. Naghavi, O. Russel, A. Etcheberry, D. Hariskos, R. Menner, M.Powalla, O. Kerrec, and D. Lincot, "The Zn(S,O,OH)/ZnMgO buffer in thin-film Cu(In,Ga)(Se,S)$_2$-based solar cells part I: Fast chemical bath deposition of Zn(S,O,OH) buffer

layers for industrial application on co-evaporated Cu(In,Ga)Se$_2$ and electrodeposited CuIn(S,Se)$_2$ solar cells," *Prog. Photovolt: Res. Appl.*, vol. 17, pp. 470–478, 2009.

[224] S. Siebentritt, "Alternative buffers for chalcopyrite solar cells," *Sol. Energy*, vol. 77, pp. 767–775, 2004.

[225] N. Naghavi, D. Abou-Ras, N. Allsop, N. Barreau, S. Bücheler, A. Ennaoui, C. Fischer, C. Guillen, D. Hariskos, J. Herrero, R. Klenk, K. Kushiya, D. Lincot, R. Menner, T. Nakada, C. Platzer-Björkman, S. Spiering, A. Tiwari, and T. Törndahl, "Buffer layers and transparent conducting oxides for chalcopyrite Cu(In,Ga)(S,Se)$_2$ based thin film photovoltaics: present status and current developments," *Prog. Photovoltaics Res. Appl.*, vol. 18, pp. 411–433, 2010.

[226] J. Bastek, N. A. Stolwijk, R. Wuerz, A. Eicke, J. Albert, and S. Sadewasser, "Zinc diffusion in polycrystalline Cu(In,Ga)Se$_2$ and single-crystal CuInSe$_2$ layers," *Appl. Phys. Lett.*, vol. 101, p. 074105, 2012.

[227] M. Contreras, B. Egaas, P. Dippo, J. Webb, J. Granata, K. Ramanathan, S. Asher, A. Swartzlander, and R. Noufi, "On the role of Na and modifications to Cu(In,Ga)Se$_2$ absorber materials using thin-MF (M=Na, K, Cs) precursor layers," in *proceedings of 26th IEEE Photovoltaic Specialists Conference, New York*, 1997, pp. 359–362.

[228] M. Ruckh, D. Schmid, M. Kaiser, R. Schäffler, T. Walter, and H. Schock, "Influence of substrates on the electrical properties of Cu(In,Ga)Se$_2$ thin films," *Sol. Energy Mater. Sol. Cells*, vol. 41-42, pp. 335–343, 1996.

[229] M. Contreras, L. Mansfield, B. Egaas, J. Li, M. Romero, R. Noufi, E. Rudiger-Voigt, and W. Manstadt, "Wide bandgap Cu(In,Ga)Se$_2$ solar cells with improved energy conversion efficiency," *Prog. Photovolt: Res. Appl.*, vol. 20, pp. 843–850, 2012.

[230] B. Canava, J. Guillemoles, E. Yousfi, P. Cowache, H. Kerber, A. Lo-effl, H. Schock, M. Powalla, D. Hariskos, and D. Lincot, "Wet treatment based interface engineering for high efficiency Cu(In,Ga)Se$_2$ solar cells," *Thin Solid Films*, vol. 361–362, pp. 187–192, 2000.

[231] M. Contreras, M. Romero, B. To, F. Hasoon, R. Noufi, S. Ward, and K. Ramanathan, "Optimization of CBD CdS process in high-efficiency Cu(In,Ga)Se$_2$-based solar cells," *Thin Solid Films*, vol. 403–404, pp. 204–211, 2002.

[232] *MultiMode^{TM} SPM Instruction Manual Verion 4.31ce*, Digital Instruments.

Appendix

Appendix A: Hardware in the SPM Control System

Connections in the circuit diagram

Connection	Description
1	Input from AFM head
2	Vertical deflection
3	Horizontal deflection
4	DC voltage V_{dc}
5	Add $V_{ac}(f_1)$ to $V_{ac}(f_0)$ (only used for frequency sweep, switched off during measurement)
6	$V_{ac}(f_1)+V_{dc}$
7	$V_{ac}(f_0)$
8	Gain select for photodiode
9	X, Y, Z via C300 DSP Controller
10, 11, 12, 13	LabVIEW communications
14	Communication with computer via LAN

Table .1: Description of the connections in the circuit diagram depicted in Fig .1.

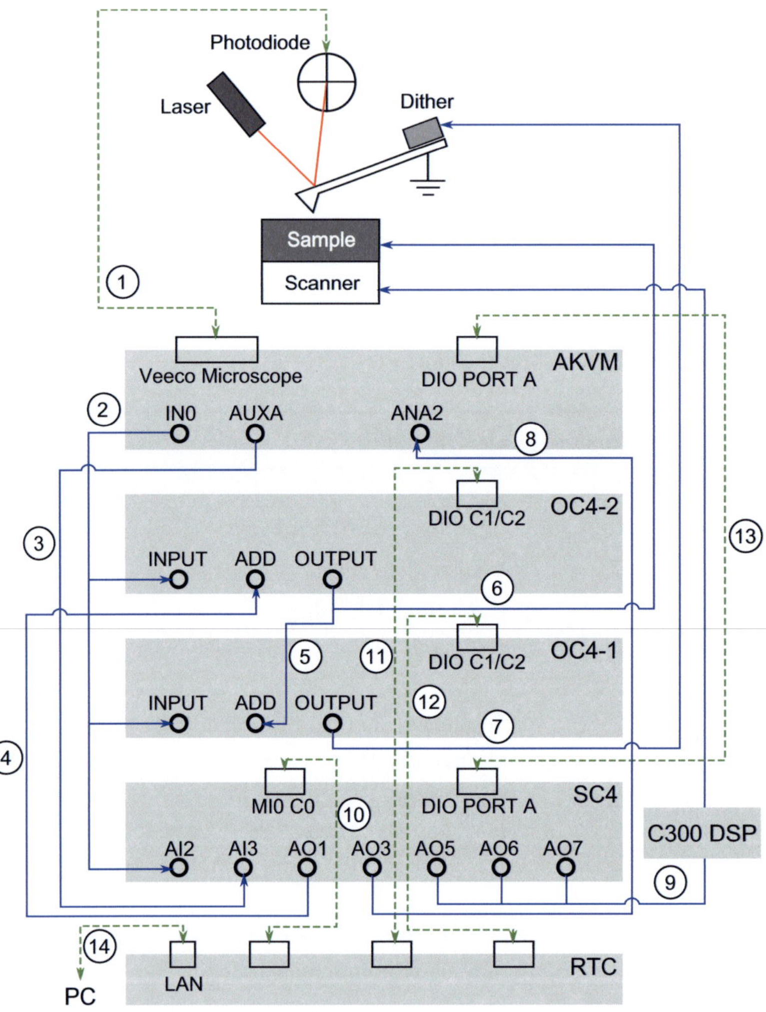

Figure .1: The circuit diagram of the Nanonis^{TM} SPM Control System with Dual-OC4. The blue solid lines with one-way arrows indicate the connections with BNC cable. The green dashed lines with two-way arrows show the communication paths via serial cable. The functionality of each connection is explained in Tab .1.

Functionalities of the control units

The AKVM is an interface between the MultiMode microscope (including the AFM head and base) and the SPM Control System. It is externally controlled via the RTC. The movement of the laser spot is input from the port Veeco Microscopy and then separately output from the ports IN0 (vertical deflection) and AUXA (horizontal deflection). The gain for the photodetector can be selected through the input ANA2. OC4-1 and OC4-2 undertake the functions of the four controllers Oscillation Control 1 (OC1), Oscillation Control 2 (OC2), Z-Controller (ZC) and Kelvin Controller (KC) illustrated in Fig. 4.1. Therein, KC is integrated in OC4-2 and the dc-voltage is applied by SC4 (from the port AO1 of SC4 to the port ADD of OC4-2). The combination of SC4 and RTC is responsible for signal conditioning and monitoring (AI2, AI3), A/D-conversion (AI2, AI3), D/A-conversion (AO1, AO3, AO5, AO6, AO7) and scan-control (AO5, AO6, AO7).

Appendix B: Calibration of the SPM Control System

Prior to each measurement the setup will be calibrated. The laser and the photodetector are adjusted until the laser beam hits the origin of the photodetector. In the OC1 panel the Drive Amplitude of $300\,\mathrm{mV}$ for $V_{ac}(f_0)$ and the central frequency f_{center} of $70\,\mathrm{kHz}$ are set for the Frequency Sweep, which is an auxiliary function of OC1. The oscillation amplitude around the given f_{center} is measured in the Frequency Sweep, so that the exact resonance frequency f_0 and Q-factor (≈ 200) of the first resonance mode can be determined. By applying f_0 the oscillation amplitude is available, which is proportional to the Drive Amplitude. For a better comparison between the measurements the Drive Amplitude is usually automatically regulated, until an oscillation amplitude of $350\,\mathrm{mV}$ is reached. Similarly, the second resonance frequency f_1 and Q-factor (≈ 550) can be determined by the Frequency Sweep in OC2. The Drive Amplitude for $V_{ac}(f_1)$ is typically set to $10\,\mathrm{mV}$ during the Frequency Sweep and $2\,\mathrm{V}$ for the measurements later on. Note that in the calibration both ac-voltages $V_{ac}(f_0)$ and $V_{ac}(f_1)$ are applied on the shaker. Therefore, the option "add" in OC1 should be turned on.

So far, the setup is calibrated without the interaction between the cantilever and the sample surface. As the next step, the cantilever is engaged until the oscillation amplitude reaches the set point of $200\,\mathrm{mV}$ in ZC. Then, a test scanning is started with SC. The topography line scan (Z) and the oscillation amplitude are monitored in the Line Scan Monitor. The parameters Proportional and Time constant in ZC are adjusted until the noise level in the topography line section and the oscillation amplitude is sufficiently suppressed. More details about the principle and the setting of these two parameters can be found in Ref. [232]. Since $V_{ac}(f_1)$ is now applied between the tip and sample, the "add" option in OC1 should be switched off and the "add" option in OC2 should be turned on. Similarly, the parameters

Proportional and Time constant in KC are adjusted according to the noise level of the Bias and X2 signals. Like in many other KPFM studies, the X2 signal instead of the oscillation amplitude is used as the error signal for the regulation [93]. The main reason is that the bias spectroscopy of the X2 signal has a zero-crossing as indicated in Fig .2, which can be regulated

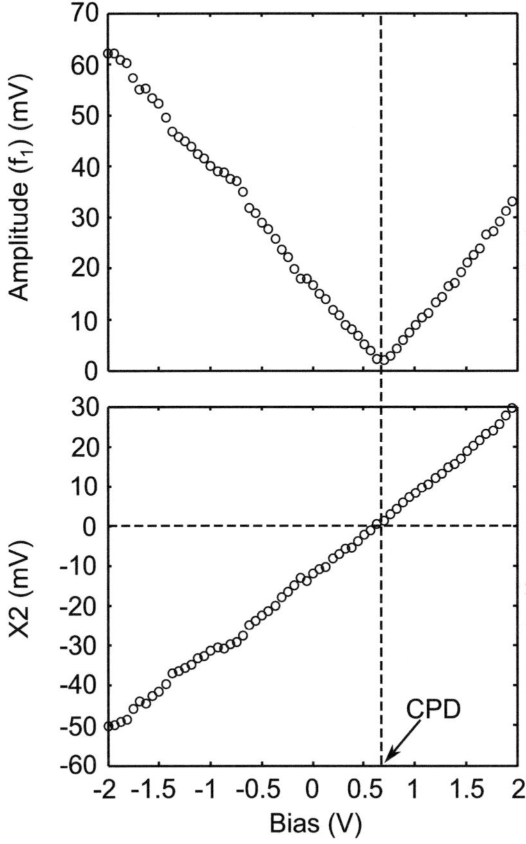

Figure .2: Bias spectroscopy of the oscillation amplitude at f_1 and its projection signal X2. The contact potential difference CPD can be determined at the bias corresponding to the minimal oscillation amplitude or the zero-crossing of the X2 signal.

with less technical difficulty. The X2 signal is in principle the projection output of the oscillation amplitude at f_1. It contains the polarity information on $V_{dc} - V_{CPD}$. Therefore, its slope varies with the reference phase set in OC2.

Finally, after the position of the scan area (Center, Size and Angle), the scan speed (Time/line) and the scan resolution (Pixels and Lines) are properly set in SC, a measurement can be routinely started.

Appendix C: Abbreviations

A_0	Oscillation amplitude at first resonance frequency
A_1	Oscillation amplitude at second resonance frequency
AAS	Atomic absorption spectroscopy
AFM	Atomic force microscopy
α	Absorption coefficient
AM	Amplitude modulation
AM 1.5	Air mass 1.5
Au	Gold
CBD	Chemical bath deposition
CdS	Cadmium sulfide
$CdSO_4$	Cadmium sulfate
CdTe	Cadmium telluride
CGS	$CuGaSe_2$, copper gallium diselenide
CIS	$CuInSe_2$, copper indium diselenide
CIGS	$Cu(In,Ga)Se_2$, copper indium gallium diselenide
CPD	Contact potential difference
Cu	Copper
$Cu(In,Ga)S_2$	Copper indium gallium disulfide

Cu_{III}	Copper on indium or gallium antisite
ΔCPD	CPD difference between ZnO and CIGS
ΔCPD_{dark}	ΔCPD in darkness
$\Delta CPD_{illumination}$	ΔCPD under illumination
E_C	Conduction band energy
$E_C(CIGS)$	Conduction band energy of zinc oxide
$E_C(ZnO)$	Conduction band energy of zinc oxide
E_F	Fermi energy
E_{Fn}	Quasi Fermi energy for electrons
E_{Fp}	Quasi Fermi energy for holes
E_g	Band gap energy
$E_g(CIGS)$	Band gap energy of copper indium gallium diselenide
EQE	External quantum efficiency
η	Power conversion efficiency
E_V	Valence band energy
$E_V(CIGS)$	Valence band energy of copper indium gallium diselenide
E_{vac}	Vacuum energy
f_0	First resonance frequency
f_1	Second resonance frequency
FM	Frequency modulation
Ga	Gallium

GaAs	Gallium arsenide
Ga_{Cu}	Gallium on copper antisite
GB	Grain boundary
GGI	[Ga]/([Ga]+[In])-ratio
HCl	Hydrogen chloride
In	Indium
In_{Cu}	Indium on copper antisite
I_{Photo}	Photocurrent
i-ZnO	Intrinsic zinc oxide
j_{sc}	Short circuit current density
jV	Current density-voltage
KC	Kelvin Controller
KPFM	Kelvin probe force microscopy
L_{eff}	Effective diffusion length
Mo	Molybdenum
MPP	Maximum power point
Na	Sodium
NH_4OH	Ammonium hydroxide
Ni/Al	Nickel-aluminium alloy
OC1	Oscillation control 1
OC2	Oscillation control 2
PE	Partial electrolyte

ϕ	Work function
Q	Quality factor
$SC(NH_2)_2$	Thiourea
SCR	Space charge region
Se	Selenium
SEM	Scanning electron microscopy
SNMS	Sputtered neutral mass spectroscopy
SPM	Scanning probe microscopy
SPV	Surface photovoltage
SSCR	Surface space charge region
STM	Scanning tunneling microscopy
V_{Cu}	Copper vacancy
V_{diff}	Diffusion voltage
V_{III}	Indium or gallium vacancy
V_{oc}	Open circuit voltage
V_{Photo}	Photovoltage
V_{Se}	Selenium vacancy
XRF	X-ray fluorescence spectroscopy
ZC	Z-Controller
$(Zn,Mg)O$	Zinc magnesium oxide
ZnO:Al	Aluminium doped zinc oxide
$Zn(OH)_2$	Zinc hydroxide

ZnS   Zinc sulfide

ZnSe   Zinc selenide

$ZnSO_4$ Zinc sulfate

Acknowledgements

Finally, I would like to express my gratitude to...

my supervisor Prof. Dr.-Ing. Michael Powalla for his continued guidance, patient advice and valuable assistance throughout this work. Without his steadfast trust, it would not have been possible for me to join Germany's first class research on CIGS.

Prof. Dr. rer. nat. Uli Lemmer for accepting me as a PhD. student at LTI and the financial support for the first two years of my research from Karlsruhe School of Optics and Photonics.

PD Dr. Hendrik Hölscher for opening up the door for me to the magic world of AFM. I would never ever forget the moments, when he helped me out with his sincere words and rich experience from the desperate situations. From him I have learned that conscientious research work can also be done in a relaxed and funny way.

PD Dr. Michael Hetterich for the patient support in the GRACIS project and his helpful physical point of views in the correction of the manuscripts.

Dr.-Ing. Alexander Colsmann and the OPV colleagues Manuel Reinhard, Hung Do, Michael Klein, Andreas Pütz, Felix Nickel, Stefan Höfle, Jens Czolk, Ralph Eckstein, as well as our industrious clean room technician Thorsten Feldmann and Christian Kayser for the nice office atmosphere. I would never forget all the events that we experienced together: the regulars' table on every Thursday, the Stadtwerke-cup, the "Canstatter Wasen", the delicious waffles, the delicious but scary "Mettfuss"...

Manuel Reinhard for the HD SEM images and the inspiring discussions all the time.

Xiaochen Tang for the excellent work during his Diploma thesis, which results in two valuable publications.

Celdal Mohan Ögün for patiently calibrating the illumination intensities of the halogen lamp.

Ute Jäntsch at IAM for supporting the wire saw, which played a decisive role in the preparation steps for untreated cross sections.

Tobias Meier and Michael Röhrig for their generous help at IMT. The scientific discussions and our social events were my most valuable memory in the huge campus north.

Dr. Wolfram Witte at ZSW for the coordination of the GRACIS project and the fabrication of the samples with varying Ga contents.

Dr. Oliver Kiowski at ZSW for the fruitful discussions in the manuscripts and the supporting of samples with ZnS buffer layer.

Dr. Erik Ahlswede, Ines Klugius, Veronika Haug, Dr. Aina Quintilla (now at KIT), Dr. Theresa Magorian Friedlmeier at ZSW for the samples prepared in the sequential process, which enabled my first talk on the international podium.

Dr. Anatoliy Slobodskyy for sharing the EQE setup at the early stage of my PhD. study.

the other LTI colleagues Patrick Schwab, Florian Maier-Flaig, Tobias Bock-srocker, Andreas Arndt, Nico Christ, Siegfried Kettlitz, Carola Moosmann for your disinterested assistance in the organization of the written exams "Solarenergie" and "Photovoltaik" in the last three years.

my parents and grandparents for their encouragement since I went to the elementary school and for backing me up in every and each of my decisions.

particularly, my wife Xue for enabling me a carefree college study and a subsequent research life. Your existence makes the life of a scientific nerd so colorful.

Thanks!

Zhenhao

Curriculum vitae

Personal information:

Name: Zhenhao Zhang

Date of birth: Sep. 20th 1984

Place of birth: Shandong/V. R. China

Education and research experience:

12. 2012   PhD. exam at Karlsruhe Institute of Technology (KIT), grade: 1.0 (Magna cum laude).

09. 2009-12. 2012 PhD. thesis on "Nanoscale investigation of potential distribution in operating $Cu(In,Ga)Se_2$ thin-film solar cells" at KIT (supervisor: Prof. Dr.-Ing. Michael Powalla).

05. 2009   Diploma exam in electrical and information engineering (Dipl.-Ing.), grade: 1.0 (Sehr gut).

11. 2008-05. 2009 Diploma thesis on "Opto-electrical investigation of hybrid solar cells based on CIGS and organic materials" (supervisor: Prof. Dr. rer. nat. Uli Lemmer).

10. 2004-05. 2009 Diploma study of electrical and information engineering at KIT.

03. 2003-03. 2004 German course, Carl Duisberg Center, Radolfzell.

09. 2002-02. 2003   Shandong University, Shandong/V. R. China.

Scholarship:

09. 2009-09. 2011   PhD. scholarship from Karlsruhe School of Optics and
Photonics (KSOP)

Industrial interships:

03. 2008-07. 2008   Alcatel-Lucent Stuttgart

08. 2006-09. 2006   Endress+Hauser Conducta Gerlingen